CAMBRIDGE LIBRARY COLLECTION

Books of enduring scholarly value

Life Sciences

Until the nineteenth century, the various subjects now known as the life sciences were regarded either as arcane studies which had little impact on ordinary daily life, or as a genteel hobby for the leisured classes. The increasing academic rigour and systematisation brought to the study of botany, zoology and other disciplines, and their adoption in university curricula, are reflected in the books reissued in this series.

Paxton's Flower Garden

Best remembered today for his innovative design for the Crystal Palace of 1851, Joseph Paxton (1803–65) was head gardener to the Duke of Devonshire at Chatsworth by the age of twenty-three, and remained involved in gardening throughout his life. Tapping in to the burgeoning interest in gardening amongst the Victorians, in 1841 he founded the periodical *The Gardener's Chronicle* with the botanist John Lindley (1799–1865), with whom he had worked on a Government report on Kew Gardens. *Paxton's Flower Garden* appeared between 1850 and 1853, following a series of plant-collecting expeditions. Only three of the planned ten volumes were published, but with hand-coloured plates (which can be viewed online alongside this reissue) and over 500 woodcuts, the work is lavish. Volume 3 includes further studies of numerous orchids, and Captain Cook's account of the discovery of the pine that would take his name, *Araucaria cookii* (Captain Cook's Pine).

Cambridge University Press has long been a pioneer in the reissuing of out-of-print titles from its own backlist, producing digital reprints of books that are still sought after by scholars and students but could not be reprinted economically using traditional technology. The Cambridge Library Collection extends this activity to a wider range of books which are still of importance to researchers and professionals, either for the source material they contain, or as landmarks in the history of their academic discipline.

Drawing from the world-renowned collections in the Cambridge University Library, and guided by the advice of experts in each subject area, Cambridge University Press is using state-of-the-art scanning machines in its own Printing House to capture the content of each book selected for inclusion. The files are processed to give a consistently clear, crisp image, and the books finished to the high quality standard for which the Press is recognised around the world. The latest print-on-demand technology ensures that the books will remain available indefinitely, and that orders for single or multiple copies can quickly be supplied.

The Cambridge Library Collection will bring back to life books of enduring scholarly value (including out-of-copyright works originally issued by other publishers) across a wide range of disciplines in the humanities and social sciences and in science and technology.

Paxton's Flower Garden

VOLUME 3

JOSEPH PAXTON
JOHN LINDLEY

CAMBRIDGE
UNIVERSITY PRESS

CAMBRIDGE UNIVERSITY PRESS

Cambridge, New York, Melbourne, Madrid, Cape Town,
Singapore, São Paolo, Delhi, Tokyo, Mexico City

Published in the United States of America by Cambridge University Press, New York

www.cambridge.org
Information on this title: www.cambridge.org/9781108037273

© in this compilation Cambridge University Press 2011

This edition first published 1852–3
This digitally printed version 2011

ISBN 978-1-108-03727-3 Paperback

PAXTON'S

FLOWER GARDEN.

———•———

BY

PROFESSOR LINDLEY AND SIR JOSEPH PAXTON.

VOL. III.

LONDON:
BRADBURY AND EVANS, 11, BOUVERIE STREET.
1852-3.

PLATE **73**

L.Constans del.&zinc.

Printed by C.F.Cheffins,London.

[PLATE 73.]

THE RETUSE ECHEVERRIA.

(ECHEVERRIA RETUSA.)

———•———

A handsome winter-flowering Greenhouse Succulent Plant, from MEXICO, *belonging to the Order of* HOUSELEEKS.

══════════

𝔖𝔭𝔢𝔠𝔦𝔣𝔦𝔠 𝔆𝔥𝔞𝔯𝔞𝔠𝔱𝔢𝔯.

THE *RETUSE* ECHEVERRIA. Caulescent. Leaves obovate, spathulate, finally scattered, glaucous, when old retuse and somewhat crenated ; those of the stem linear-oblong, entire, free at the base. Panicle small, dense, divaricating, somewhat corymbose, with few-flowered branches. Sepals narrowly ovate, acute, unequal, much shorter than the corolla. Petals acute, keeled, gibbous at the base.

ECHEVERRIA *RETUSA ;* caulescens, foliis obovato-spathulatis demum sparsis glaucis ; vetustis retusis crenulatis ; caulinis lineari-oblongis integerrimis basi solutis, paniculâ parvâ densâ divaricatâ sub-corymbosâ ramis paucifloris, sepalis angustè ovatis acutis inæqualibus corollâ multò brevioribus, petalis carinatis acutis basi gibbosis.

Echeverria retusa : *Lindley, in Journ. of the Hort. Soc.,* vol. ii., p. 306.

══════════

THIS is by no means so well-known a plant as its usefulness should have rendered it, seeing that it was published almost five years since in the *Journal of the Horticultural Society,* with the following account :—

"It was raised from seeds, received from Mr. Hartweg in February, 1846, and said to have been collected on rocks near Anganguco, in Mexico. A dwarf species, not unlike a contracted form of *E. Scheerii.* Its leaves are originally closely imbricated, but are never truly rosulate, and by degrees separate as the stem lengthens ; they are broad at the point, but acute when young ; when old become extremely blunt, and irregularly crenated, as well as bordered with purple. The flower-stem is from nine inches to more than a foot high, and bears at the very summit a compact panicle of handsome crimson flowers, covered with a delicate bloom, and orange-coloured inside. It is a pretty

B

greenhouse, half-shrubby species, growing from one to two feet high, and thriving vigorously in a light mixture of sandy loam with leaf-mould and plenty of sand. It is easily increased by the leaves, and flowers freely from November to April, that is to say, throughout the winter."

No plants are better suited to window gardens than these Echeverrias, all the species of which blossom the whole winter long, will thrive in soil of any sort, are not very impatient of either heat or cold, dryness or dampness, and which are so varied in colour, form, and manner of growth, as to form of themselves variety enough for such a space as the recess of a window affords. One of the most singular is the *Pachyphytum bracteosum* of Klotzsch, which does not appear to be in any way distinguishable from the genus.

PLATE 74.

L.Constans del. & zinc.

Printed by C.F.Cheffins, London.

[Plate 74.]

THE THYRSE-LIKE BILLBERGIA.

(BILLBERGIA THYRSOIDEA.)

———————◆———————

A Stove Perennial, from BRAZIL, *with rich crimson bracts, arranged in a cone, belonging to* BROMELIADS.

════════════════

Specific Character.

THE THYRSE-LIKE BILLBERGIA. Leaves erect, broadly strap-shaped, obtuse with a point, uniformly concave, spiny-toothed, about as long as the scape. Bracts ovato-lanceolate, acuminate, collected into a cone or spike like a thyrse in form. Calyx covered with white mealiness. Petals obtuse, much longer than the calyx.

BILLBERGIA *THYRSOIDEA ;* foliis erectis lato-ligulatis obtusis cum acumine æqualiter concavis, spinoso-serratis scapo sub-æqualibus, bracteis ovato-lanceolatis acuminatis in strobilum aggregatis, spicâ thyrsoideâ, ovariis albo-farinosis, petalis obtusis calyce multò longioribus.

Billbergia thyrsoidea : *Martius in Römer and Schultes Sp. Plant.,* 7., 1261.

════════════════

A PLANT with the manner of growth and appearance of a Pine Apple, except that the leaves are wholly destitute of a mealy or glaucous covering, but are a clear bright green. Leaves loosely arranged, rather wavy, with small prickly serratures, and a short abrupt point. Bracts rich crimson, very regularly arranged in an oblong obtuse cone, or thyrse, not mealy. Flowers rather larger than the bracts, and of nearly the same colour. Sepals oblong, obtuse, smooth, much shorter than the closed-up straight erect petals. Stamens six ; three free, and opposite the sepals ; three united to about the middle of the petals which have at the base a pair of half ovate scales, the outer edge of which is coarsely toothed. Ovary covered with a fine white loose mealiness, which is composed of minute oval loose cells filled with air ; three-celled, with numerous anatropal ovules having an elevated raphe, a crested chalaza, and a large secundine projecting beyond the orifice of the primine ; the stigmas are three, and convolute.

Such are the characteristic marks of this very beautiful stove plant, originally found by Martius on rocks near Rio Janeiro, and recently imported by M. de Jonghe of Brussels. For the opportunity of figuring it we are indebted to Mr. Henderson of the Wellington Nursery, St. John's Wood. It requires to be managed in the same way as a Pine Apple.

It is most nearly allied to the Pyramidal Billbergia figured in the *Botanical Magazine,* t. 1732, and in the *Botanical Register,* t. 203 and 1181; but that plant has glaucous taper-pointed leaves, and very large spreading flowers, conspicuous for the white mealiness of the calyx.

PLATE **75**

I.Constans del.& zinc.

Printed by C.F.Cheffins,London.

[PLATE 75.]

THE GOLDEN SWAN-ORCHIS.

(CYCNOCHES AUREUM.)

A noble Epiphyte, with clear yellow flowers, from CENTRAL AMERICA, *belonging to* ORCHIDS.

Specific Character.

THE GOLDEN SWAN-ORCHIS. Raceme long, pendulous, compact. Sepals lanceolate, flat. Petals of the same form, but rolled backwards from the point. Lip with a short stalk, at the end ovate and acute, with a round disk the edge of which is broken up into short curved processes forked at the point ; the two lowest larger, distinct, and straight. Column the length of the lip.

CYCNOCHES *AUREUM ;* racemo longo pendulo compacto, sepalis lanceolatis planis, petalis conformibus ab apice revolutis, labello brevi-unguiculato apice ovato acuto, disci rotundati margine in processubus brevibus arcuatis apice furcatis soluto : 2 basilaribus majoribus discretis rectis, columnâ labelli longitudine.

To the very singular race of Swan-Orchises, we have now the gratification of adding a new form, introduced from Central America by Mr. Skinner. It is very near the " Spotted," from which it differs in having a shorter and more compact raceme, whole-coloured pale clear yellow flowers, and a lip the terminal lobe of which is short and ovate, not long and linear-lanceolate, while the appendages into which the edge of the disk is broken up are short, forked, all radiating from the centre, instead of the uppermost one being bent back, and the two lowest are very considerably larger than the others.

Is this a species ? or is it a form of *C. maculatum,* or of some other of this masquerading genus ? Upon this subject we venture to repeat what was said six years ago in the *Botanical Register,* upon the surprising transformations to which the Swan-Orchises are subject, and concerning which we have no more information than we had in 1846. The plant to which the remarks applied was the green state of the Egertonian Swan-Orchis.

"This is evidently a variety of the *C. Egertonianum,* distinguished by its flowers being of a pale watery green, and not deep purple. But what is *C. Egertonianum* itself ? In Mr. Bateman's magnificent work we are told how the long-spiked small purple-flowered *C. Egertonianum* is only the short-spiked large green-flowered *C. ventricosum;* how the same plant at one time bears one sort of flowers, and at another time another sort; and we have ourselves shown how the same plant, nay the same spike, is sometimes both the one, the other, and neither. *C. Egertonianum* is then a ' sport,' as gardeners say, of *C. ventricosum.*

" But what again is *C. ventricosum?* Who knows that it is not another 'sport' of *C. Loddigesii,* which has indeed been caught in the very act of showing a false countenance, something wonderfully suspicious, all things considered, and justifying the idea that it is itself a mere Janus, whose face is green and short on one side, and spotted and long on the other.

" Then, if such apparently honest species as *C. Egertonianum, ventricosum,* and *Loddigesii* are but counterfeits, what warrant have we for regarding the other so-called species as not being further examples of plants in masquerade? For ourselves we cannot answer the question: nor should we be astonished at finding some day a Cycnoches no longer a Cycnoches, but something else; perhaps a Catasetum. If one could accept the doctrine of the author of the 'Vestiges,' it might be said that in this place we have found plants actually undergoing the changes which he assumes to be in progress throughout nature, and that they are thus subject to the most startling conditions only because their new forms have not yet acquired stability."

Since we have space for the purpose, we avail ourselves of the opportunity to give a list of the known forms of this strange genus.

SO-CALLED SPECIES OF CYCNOCHES.

* *Lip perfectly entire, fleshy, without appendages.*

1. C. Loddigesii *Lindl. Gen. & Sp. Orch.*, p. 154 ; *Bot. Cab.*, t. 2000 ; *Bot. Reg.*, t. 1742.—*Surinam.*—Flowers very large, fragrant, green and purple, with a white spotted lip. Sports by producing smaller broad-lipped flowers without scent, and with a very short cucullate club-shaped column. This is the original state of the genus.

2. C. ventricosum *Bateman Orch. Mex. & Guatemala,* t. 5.—*Guatemala.*—Flowers large, green, with a white lip. Sports to *Egertonianum;* and even towards the cucullate form of *C. Loddigesii,* as was ascertained by Sir P. Egerton, in 1849.

3. C. chlorochilon *Klotzsch; Sertum Orchidaceum,* t. 16.—*Maracaybo.*—Flowers very large, green, whole-coloured. Has not been observed to sport ; but is probably a mere variety of *C. ventricosum.*

* * *Lip having the edge broken up into fleshy appendages.*

4. C. pentadactylon *Lindl. in Bot. Reg.,* 1843, misc. 26, t. 22.—*Brazil.*—Flowers large, yellowish green, banded with brown. In the garden of Mr. Kenrick, of West Bromwich, this produced two flowers of *Egertonianum,* among the usual flowers peculiar to itself, Sept. 12, 1851.

5. C. aureum *Lindl. in Paxt. Fl. Garden,* t. 75.—*Central America.*—Flowers large, clear pale yellow. Has not been yet observed to sport.

6. C. maculatum *Lindl. in Bot. Reg.,* 1840, misc. 8 ; *Sertum Orchidaceum,* t. 33.—*Mexico? La Guayra.*—Flowers small, yellow, spotted with brown. Has not been observed to sport.

7. C. Egertonianum *Bateman Orch. Mex. & Guatemala,* t. 40 ; *Bot. Reg.,* 1846, t. 46.—*Guatemala and Mexico.*—Flowers small, purple or greenish, unspotted. Sports to *Ventricosum,* and to *Pentadactylon.*

* * * *Lip three-lobed, membranous, without appendages.*

8. C. Pescatorei *Lindl. in Paxt. Fl. Gard.,* no. 174 ; *aliàs* Acineta glauca *Linden.*—*New Grenada.*—Flowers yellow and brown, in a long pendulous raceme. Has not been observed to sport.

9. C. barbatum *Lindl. in Journ. of Hort. Soc.,* vol. iv. ; *Bot. Mag.,* t. 4479.—*New Grenada,* and *Costa Rica.*—Flowers soft delicate flesh-colour, spotted with red. Has not been seen to sport.

GLEANINGS AND ORIGINAL MEMORANDA.

464. CHÆNOSTOMA LINIFOLIUM. *Bentham.* (*aliàs* Manulea linifolia *Thunberg; aliàs* Chænostoma fasciculatum *of Gardens.*) A beautiful little shrub, with long white flowers having a yellow orifice. Belongs to Linariads. Native of the Cape of Good Hope. (Fig. 233.)

We think there can be no doubt that the *Ch. fasciculatum* of Gardens is identical with *Ch. linifolium*, notwithstanding that its flowers are much longer and more loosely arranged than they are found in the stunted specimens preserved in herbaria. It may be regarded as a form of that plant, with all the parts drawn out by exuberant growth. It forms a neat, dwarf, compact bush, with narrow leaves, which are sometimes bluntly toothed, and long loose racemes of tubular white flowers, orange-yellow at the mouth, beyond which the yellow anthers project. It blossoms late in the autumn, or early in winter, according to the treatment it receives. A mixture of peat, loam, and sand suits it perfectly. When out of flower, it should be allowed to complete its growth, and then be rested for three or four months. It must have abundance of air at all times. Cuttings multiply it readily. It may also be treated like a tender annual ; in this respect resembling such plants as Mignonette, which are really undershrubs, although flowering the first year.

465. CALODRACON NOBILIS. *Planchon.* (*aliàs* Calodracon Sieboldii *Planchon; aliàs* Dracæna nobilis *Van Houtte.*) A hothouse plant with a graceful but noble habit, and rich purple and crimson leaves. Native of Japan. Belongs to Lilyworts.

This plant, already known in gardens under the name of *Dracæna nobilis,* resembles the *Calodracon Jacquini* of Planchon (*Drac. ferrea* and *terminalis* of books), and is remarkable for the singularly vivid mixture of streaks of rich crimson and purple in its foliage. It is said to be more dwarf than the last species, and far more attractive. " Entre mille plantes d'une serre, c'est sur elle que se portent d'abord les regards ; dans un salon, c'est l'ornement le plus exquis que la nature puisse prêter au raffinement de luxe ;" such is the flowery language in which M. Planchon speaks of it in Van Houtte's *Flore des Serres*, where there is an excellent figure of the species. It has not yet flowered. The stem is described as being so short as to be almost concealed by the head of leaves ; nothing, it is added, can be more beautiful either in the stove itself, or in a vase in a sitting-room warm enough to keep it in health, and sufficiently lighted.

466. COMMELYNA SCABRA. *Bentham.* A half-hardy perennial plant, with glaucous wavy leaves, and large dull purplish brown flowers. Native of Mexico. Belongs to the Order of Spiderworts. Introduced by M. Allardt of Berlin. (Fig. 234.)

A very singular herbaceous plant, first found by Mr. Ehrenberz in the North of Mexico, and afterwards by Hartweg. It forms a tuft of straggling stems variegated with red. The leaves are sessile, lanceolate, stiff, cartilaginous at the edge, covered all over with fine asperities, with purplish sheaths fringed at the orifice. The spathes are almost cordate, folded together, downy, with five to ten flowers in each. The petals are of a singularly dull purplish brown colour.—*Link, Klotzsch and Otto, Icones*, t. 30. This does very well in a warm border out of doors in the summer, but as it dislikes wet and cold, its roots must be taken up in the autumn and kept dry over the winter. It requires a light rich garden soil.

467. GRINDELIA GRANDIFLORA. *Hooker.* A hardy biennial, with large showy orange-coloured flower-heads. Native of Texas. Belongs to Composites. Introduced at Kew.

Raised from seeds sent by Dr. Wright from Texas, and quite hardy, flowering in the open air as late as November 1st, when our drawing was made. In foliage the species certainly more closely resembles *G. inuloides*, Bot. Reg. t. 243, than *G. squarrosa*, figured in *Botanical Magazine*, t. 1706, but it appears on comparison distinct from both, especially in the great size of the flowers (capitula) and in the deep orange-yellow of the broad ray, no less than in the great height of the plant, three to five feet in our garden. It must be confessed, however, that the species of the genus are very variable and ill-defined. Stems, on an average, four feet high, erect, herbaceous, simple till towards the summit, where they are corymbosely branched, each branch leafy and terminated by a flower. Whole plant hard and rigid, sub-glaucous. Leaves alternate, sessile, from a broad cordato-semiamplexicaul base, lanceolate, gradually tapering to a point ; the base coarsely dentato-serrate, the rest nearly entire. Flowers (capitula) very large, solitary, on each terminal branch, full orange-yellow. Involucre hemispherical, glutinous : scales subulate, spreading or even recurved, squarrose, herbaceous. Radical florets ligulate, very long, with a slender tubular base. Ovary obovate, furrowed, bearing one or more setæ : style with the branches subulate. Florets of the disc tubular, five-toothed, of the ovary, as in the ray, setæ three to six. A stout

234

plant, making a showy appearance when in flower. Towards autumn the stem becomes hard and woody ; after flowering, the stem and roots are exhausted and die, showing that the plant is only a biennial. Like many Mexican Compositæ, it does not freely ripen seeds ; but it may be readily increased by cuttings, which should be struck so as to have them established by the end of the summer, the young plants being kept in a cool airy place till the spring, when they may be planted out in the flower-borders.—*Bot. Mag.*, t. 4628.

468. ODONTOGLOSSUM ANCEPS. *Klotzsch.* A diminutive epiphyte with greenish yellow flowers, and a white lip. Native of Brazil. Flowers in July. Introduced by M. Allardt of Brazil.

O. anceps ; pseudobulbis compressis, versus apicem attenuatis ; foliis binis, oblongis, apice oblique rotundatis, ternis ; racemo ancipiti, unifloro, foliis longiore, biarticulato ; vagina bivalvi, membranacea, subarida ; perigonii foliolis lanceolato-oblongis, obtusis, patenti-recurvis ; labello rhomboideo-lyrato, apice recurvo, appendiceque bidentato atque anteriore intermedio breviore instructo.

Pseudobulbs two inches long. Leaves four inches long, by six lines broad. Scape two-edged, four and a half inches long. Sepals an inch long, greenish yellow, the two side ones with a purple line, the upper and the petals without marks. Lip white, with purple lines and spots at the base.—*Klotzsch, in Allgem. Gartenzeit., Aug. 9, 1851.*

469. CALCEOLARIA STRICTA. *Humboldt & Bonpland.* A handsome half-hardy shrub from Peru. Flowers pale yellow, appearing in September. Introduced by Messrs. Veitch & Co. (Fig. 235.)

This is another of those valuable, shrubby, willow-leaved Calceolarias which, independently of their intrinsic merit, will become so important as breeders. It is nearly related to the *C. tetragona* mentioned by us under No. 337, fig. 170 ; but differs in the form of the leaves, the size and colour of the flowers, and the proportion of the calyx. This species forms a small smooth bush, with willow-like leaves, pallid beneath, finely toothletted on the edge. The flowers are pale yellow, with the upper lip of the corolla much smaller than the lower, and rather distinctly crenated at the angle where its edges curve inwards. The calyx is very much shorter. It seems to be a very common Peruvian plant. We saw it last October growing in Messrs. Veitch's nursery in the open air, and flowering in great beauty. No doubt it should be planted out in the summer, in light friable soil, and removed to a conservatory in winter. Messrs. Veitch inform us that it lived through the winter of 1850-1 in a cold frame. Mr. Wm. Lobb, who found it near Loxa, describes it as a shrub from two to three feet high.

235

470. IMPATIENS CORNIGERA. *Hooker.* A robust and handsome tender annual, with clusters of hairy large purple and yellow flowers in the axils of the leaves. Native of Ceylon. Introduced in 1851.

Raised in the stove of the Royal Gardens, from seeds sent from Ceylon by Mr. Thwaites. It flowered the whole summer and autumn, and may be pronounced a really ornamental plant. In our Herbarium we find specimens which we consider to be identical, from Assam, sent by Major Jenkins, and among those specimens are some with glabrous flowers, which have considerable affinity with *Impatiens lævigata* Wall., but from which the present appears truly distinct. Stem erect, three to four feet high, rather stout, succulent, semipellucid, striated, often red at the setting on of the leaves, very thick and much branched and rooting below. Leaves alternate, large, sometimes nearly a span long, ovate, acuminated, penninerved, pale beneath ; petiole and midrib generally red, the margin very obscurely crenato-serrate, the minute teeth bearing a seta which is long and conspicuous at the base of the leaf ; the edge too, as seen under a lens, is everywhere ciliated; petiole one to two inches long, and nearly a line broad, semiterete, margined, the margin bent, with more or less numerous long, soft, distant fimbriæ tipped with a gland. Peduncles aggregate, axillary, single-flowered, much shorter than the petiole, a little enlarged upwards, and curved down with the weight of the flower. The size of the flower is about equal to those of *Impatiens balsamina*, and the colour is yellowish, much suffused with pink. The upper sepal (two united) is remarkable for a large green horn-like projection from the back ; the lower for being downy, and for the short, much-curved, green spur. This, like other tropical species of the genus, requires to be treated as a tender annual. If potted in light rich soil, and kept in a stove and well supplied with water, it attains a considerable size, producing thick side-branches, which in time assume a hard woody appearance. When placed in a favourable situation as regards shade and moisture, the lower parts of the branches produce aerial roots, which descend till they reach the soil, and then materially assist in supplying nourishment to the plant. As it flowers late, we fear it will not ripen seeds ; but it may be increased by cuttings, which root readily in the summer, but require much care in the winter, as they are liable to damp off.—*Bot. Mag.,* t. 4623.

471. SOPHRONITE, THE SPECIES OF.

The Sophronites form a very distinct little genus, all the species of which are gems nestling in moss, upon the branches of old trees in Brazil. With the exception of *S. cernua* they are little known, and therefore a short history of them, illustrated by the accompanying woodcut, may be useful both to cultivators and botanists. The genus was first proposed at fol. 1129 of the *Botanical Register,* under the name of Sophronia, afterwards at t. 1147 of the same work changed to Sophronitis. The original species named *S. cernua,* imported from Botofogo, a place in the neighbourhood of Rio Janeiro, was for a long time the only kind known in gardens, and appears to have since given rise to three other names, viz., *S. isopetala, Hoffmannseggii,* and *nutans,* the plants bearing which are not in any way distinguishable by the accounts their authors have published of them. A second species was added in the *Sertum Orchidaceum,* with large scarlet flowers, under the name of *S. grandiflora;* then in 1840 came a third with violet flowers, called *S. violacea;* and a fourth, *S. pterocarpa,* has long lain buried in herbaria. A good generic character not having been yet published, we offer the following as one applicable to all the four species now known :—

Perianthium explanatum, subæquale. *Sepala* et *Petala* imbricata, libera. *Labellum* integrum, cucullatum, linguiforme, basi cum columna connatum, sæpiùs cristâ simplici transversâ in medio lamellisque 2 axialibus. *Columna* libera, apice utrinque alata : alis integris conniventibus super cristam labelli. *Stigma* concavum, rostello obtuso. *Anthera* terminalis, opercularis, 8-locularis, cardine crasso inarticulato. Pollinia 8, anticè et posticè parallela, caudiculâ duplici pulvereâ. —Herbæ *epiphytæ* (Brasilienses) *monophyllæ, pseudobulbosæ, racemis axillaribus effusis paucifloris,* floribus coccineis v. violaceis.

Of this the following are the species with their distinctive characters :—

472. SOPHRONITIS CERNUA *Lindley in Botanical Register,* t. 1129 ; (*aliàs* S. isopetala *Hoffmannsegg in Botan. Zeitung,* I. 834 ; *aliàs* S. Hoffmannseggii *Reichenbach fil in Linnæa Litt. Ber.,* XVI. 236 ; *aliàs* S. nutans *Id. Ibid.;*) folio ovato-oblongo, racemo corymboso paucifloro, sepalis petalisque ovatis acutis, labello repando acuto, columnæ alis brevibus obtusissimis, ovario sexcostato. (Fig. 236 ; 8, a lip ; 9, pollen masses ; 10, an end view of the ovary.)

This plant has small brilliant scarlet flowers, with a yellow lip. The sepals and petals are of the same size. There does not seem to be any essential difference in the plants now referred here. The species is common in gardens.

473. SOPHRONITIS GRANDIFLORA *Lindley Sertum Orchidaceum,* t. 5, fig. 2 ; (*aliàs* Cattleya coccinea *Bot. Reg.,* fol. 1919;) folio oblongo acuto pseudobulbo ovali tereti longiore, floribus solitariis, spathâ nullâ, sepalis lineari-oblongis obtusis rectis, petalis triplò latioribus, labello ovato basi cucullato indiviso apice plano sepalis breviore. (Fig. 237.)

Found by Descourtilz, upon the high mountains that separate the province of Bananal from that of Ilha Grande ; by Gardner, on trees near Rio Janeiro, on mountain heights, where rime frost is seen in the morning (659 and 5878 of his Herbarium). The finest of the genus. Flowers bright scarlet or cinnabar, three inches across ; lip yellow.

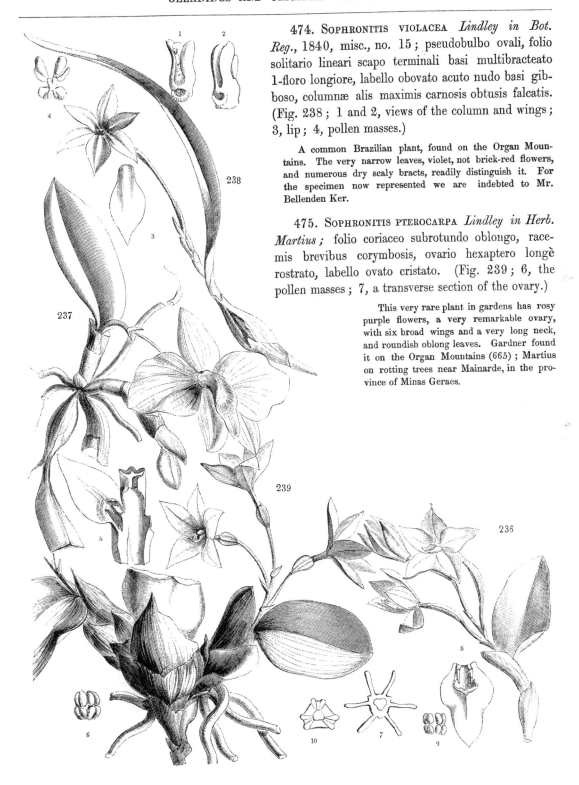

474. SOPHRONITIS VIOLACEA *Lindley in Bot. Reg.*, 1840, misc., no. 15; pseudobulbo ovali, folio solitario lineari scapo terminali basi multibracteato 1-floro longiore, labello obovato acuto nudo basi gibboso, columnæ alis maximis carnosis obtusis falcatis. (Fig. 238; 1 and 2, views of the column and wings; 3, lip; 4, pollen masses.)

A common Brazilian plant, found on the Organ Mountains. The very narrow leaves, violet, not brick-red flowers, and numerous dry scaly bracts, readily distinguish it. For the specimen now represented we are indebted to Mr. Bellenden Ker.

475. SOPHRONITIS PTEROCARPA *Lindley in Herb. Martius;* folio coriaceo subrotundo oblongo, racemis brevibus corymbosis, ovario hexaptero longè rostrato, labello ovato cristato. (Fig. 239; 6, the pollen masses; 7, a transverse section of the ovary.)

This very rare plant in gardens has rosy purple flowers, a very remarkable ovary, with six broad wings and a very long neck, and roundish oblong leaves. Gardner found it on the Organ Mountains (665); Martius on rotting trees near Mainarde, in the province of Minas Geraes.

476. SKIMMIA JAPONICA. *Suprà*, vol. ii., no. 318, fig. 163.

At the above place we referred to this plant, as a synonyme, the *Limonia Laureola* of Wallich, the materials at our disposal not enabling us to point out any difference. We have since been supplied with further information by Messrs. Standish and Noble, who have fruited the *Skimmia japonica* in abundance, and we are now satisfied that the two plants are distinct. The following letter from Mr. Standish includes the main points of difference :—

"Sir,—At your request, I have much pleasure in sending two or three seeds of *Skimmia japonica*. They have been gathered and put in sand more than a month—therefore are almost spoiled for your purpose. When perfect, they contain two seeds, but never more. Amongst the quantity that we have pulled to pieces for sowing, about one third contained two seeds—the rest only one. The whole of the berries were oval in shape. Enclosed is a leaf of our plant, and one from Mr. Luscombe's plant (*Limonia Laureola*). If you bruise the two you will find a great difference in the scent. Ours flowers at two inches high, and fruits at six inches ; the other, although a very large plant, has never flowered out or fruited. M. Van Geert, of Belgium, tells me that he has had *Limonia Laureola* many years—is quite satisfied it is not the same as ours ; and, although his plant is three feet in height, and every year has all the appearance of coming into bloom, yet never comes. Many persons are selling *Limonia Laureola* for *Skimmia japonica*, and the public will be very much disappointed when they come to see the two plants, therefore I think they ought to be made acquainted with these facts. Every one who has seen *Skimmia japonica* in fruit, has been charmed with it. We find it perfectly hardy ; and, whether looking at it as an evergreen, or its very sweet-scented flowers or fruit, it is a very fine plant."

These statements we can quite confirm ; for, although both have sweet-scented leaves, yet *Limonia Laureola* is by no means so sweet as *Skimmia japonica*. The form of the leaves, too, is different—in the former flat and nearly acute—in the other more lanceolate, rather wavy, and acuminate. We must, however, add that the statement of the authors of the *Flora Japonica*, that the seeds of the *Skimmia japonica* have no albumen, is undoubtedly a mistake. We find in Mr. Standish's perfectly ripe seeds, a large greenish embryo, with a thick layer of white albumen between it and the skin. In this respect then, the two plants are alike, and the supposed difference, of albumen in *Limonia Laureola*, and none in *Skimmia japonica*, falls to the ground.

To prevent further error, we put the distinctions of the two species into the following technical form :—

1. *S. japonica* (Thunberg, and our fig. 163) ; foliis lanceolatis acuminatis undulatis pyri olentibus.
2. *S. Laureola* (*aliàs* Limonia Laureola *Wallich*) ; foliis oblongis acutis planis rutæ olentibus.

The scent of the leaves of the first seems to us to resemble ripe apples, of the latter a mixture of Rue and Fraxinella.

477. MAXILLARIA PUNCTULATA. *Klotzsch.* A Brazilian epiphyte with greenish yellow flowers, and a three-lobed yellowish lip spotted with purple at the edge. Introduced to the German gardens.

M. punctulata ; caulescens ; pseudobulbis oblongis, versus apicem attenuatis bifoliatis, bifariam imbricatis ; foliis coriaceis unicostatis, ligulatis, acutis, subtortuosis, basi conduplicato-attenuatis ; pedunculis unifloris, foliis brevioribus, teretibus 4—5-bracteatis ; bracteis amplexicaulibus, subcarinatis, membranaceis, a germine remotis ; perianthii foliolis oblongis, lanceolatis, acutis, exterioribus patentibus, interioribus, duplò minoribus conniventibus ; labello oblongo, trilobo, centro lævi, supra basin subpulvinato, lateralibus brevibus, intermedio obtuso, subrecurvo ; anthera puberula.

Pseudobulbs three inches long, somewhat compressed, eight lines wide. Leaves from six to seven inches long, twelve to fourteen lines wide. Peduncles six inches long. Sepals an inch long, three lines broad. Petals eight lines long, and one and a half broad. Lip eight lines long. M. L. Mathieu, who flowered this plant in July, 1851, received it from M. Linau of Frankfort on the Oder, to whom it was sent direct from Rio Janeiro in Brazil by his nephew.— *Klotzsch, in Allgem. Gartenzeit., Aug. 9, 1851.*

478. EPIDENDRUM WAGENERI. *Klotzsch.* An orchidaceous epiphyte with greenish yellow flowers, and a white lip streaked with purple. Native of Venezuela. Introduced at the Botanic Garden, Berlin.

E. Wageneri (Encyclium) ; pseudobulbis cæspitosis, ovatis, 2—3-foliatis ; foliis linearibus, unicostatis, coriaceis, tortuosis, apice obtusis ; racemo paniculato terminali, viridi-punctulato ; germinibus teretibus, punctato-scabris ; perigonii foliolis æqualibus, spathulato-acutis, patentibus ; labello trilobo candido, lineis elevatis striato, basi bicalloso, lobis lateralibus brevibus, conniventibus, falcato-oblongis, obtusis, lobo intermedio cordato, orbiculato-ovato, brevi-acuto ; columna libera apice utrinque uncinato-auriculata.

Pseudobulbs two inches long, and one and a half in diameter above the base. Leaves fifteen inches long, and eight lines wide. Racemes two feet long. Flowers sweet-scented, yellowish green. Lip white, changing to clay-colour.— *Klotzsch, in Allgem. Gartenzeit., Aug. 9, 1851.*

479. EPIDENDRUM COLORANS. *Klotzsch.* An orchidaceous epiphyte with small pink flowers. Native of Guatemala. Introduced by Mr. Warczewicz. Flowered with M. Allardt of Berlin.

E. colorans (Spathium) ; caule tereti ; foliis distichis, patentirecurvis, oblongis, brevi-acutis, subcarinatis ; racemo brevi, punctato-scabro ; spatha, dorso crenulata, diphylla ; floribus parvis, brevi-pedicellatis, bractea lanceolata, acuminata, pallide lilacina suffultis ; perigonii foliolis spathulatis, interioribus subbrevioribus angustioribusque ; labelli trilobi lobis lateralibus brevioribus, integerrimis, subobliquis, intermedio elongato, apice bilobo, lamellis 3 parallelis, angustis instructo.

Stems a foot high. Flowers pendulous small, bracts two lines long. Sepals white at first, then pink; petals narrower and a little shorter. Lip white with three narrow plates.—*Klotzsch, in Allgem. Gartenzeit., Aug. 9*, 1851.

480. ILEX LATIFOLIA. A hardy evergreen tree, with long shining leaves, greenish flowers, and small red axillary berries. Said to be a native of Japan. Belongs to Aquifoils. (Fig 240.)

This is a stout, stiff, evergreen, hardy tree, of great beauty. Every part is entirely free from hair. The shoots, which

240

are deep green or tinged with violet, are somewhat angular near the ends. The leaves, which are from six to eight inches long, are deep green, not coloured at the edge, flat, oblong, acuminate, sharply and pretty regularly serrated, except at the base, which is entire, and gradually narrows into a petiole about three quarters of an inch long. The flowers are small, hermaphrodite, pale green, in very close axillary racemes, about as long as the leafstalks, and supported by short, ovate, acute, shining, cartilaginous bracts. The berries, which ripen in February, are in short compact clusters, of a dull red colour, and nearly spherical; each contains from four to five stones, in which we have never succeeded in finding a kernel.

This valuable plant passes under the name of *Ilex latifolia*, by which Thunberg designated a small tree called, in Japan, *No-Ko-Giri*; but, if the statement of that botanist can be trusted, his plant must be different, for he says the leaves are egg-shaped, and three inches long by two broad, which gives them an entirely different outline from the species before us, the proportion of whose leaves is not three by two, but six or seven by two, a very material difference. Nevertheless, in the absence of any authentic evidence, we leave the garden name as we find it, especially since it is probably the *I. latifolia* of Zuccarini and Siebold (*Floræ japonicæ familiæ naturales*, sect. i., p. 40), or *I. macrophylla* of Blume. According to the first of these authors, the leaves in the wild plant vary in form, being, on the same branch, oblong, ovate, or elliptical, acuminate or obtuse, and finely serrated, or slightly crenate.

The species nearly approaches the *Ilex Perado* of the Hortus Kewensis, a native of the Canaries, figured in the *Botanical Magazine*, t. 4079, under Webb and Berthellot's name of *I. platyphylla*, another very handsome hardy shrub, differing from this in bearing clusters of large white flowers, and fruit more than twice the size of that of the present plant. There is no doubt that this *I. latifolia*, of which we believe two varieties are in cultivation, and which is plentiful in the nurseries, is as hardy as the common holly itself.

481. EUGENIA UGNI. *Hooker.* (*aliàs* Myrtus Ugni *Molina; aliàs* Murtilla *Feuillée.*) A beautiful evergreen bush, with globular pink and white flowers, and fragrant foliage. Belongs to Myrtleblooms. Native of Chili. Introduced by Messrs. Veitch.

It forms a charming shrub, native of South Chili and the islands, abundant in Chiloe and in the Bay of Valdivia, where the natives call it *Ugni*, and the Spaniards *Murtilla* or *Myrtilla*; and the habit is not unlike that of our European Myrtle. Introduced by Messrs. Veitch and Son, through their collector, Mr. William Lobb. It proves quite hardy in their Nursery at Exeter, whence we were favoured with the flowering specimen here figured in July, 1851. The flowers are fragrant, and the leaves when bruised are no less so; which ensures its being prized by all cultivators. A shrub, varying in height, according to Mr. Bridges, from two to four feet, copiously branched; branches erecto-patent, clothed with brown bark, young shoots downy. Leaves copious, opposite, spreading, on very short petioles, thick, coriaceous, ovate, sometimes varying to lanceolate, very acute, impunctate, nerveless, the margin reflexed, dark green above, pale and when dry almost white beneath. Peduncles axillary, solitary, single-flowered, with a pair of linear reflexed bracts at the setting on of the flower. Calyx-tube turbinate, dotted : limb of five (or rarely four) recurved, linear lobes, exactly resembling the bracts. Petals five (or four), erect, orbicular, very concave (forming a globose corolla), white, tinged with rose. Stamens numerous : anthers red. Style shorter than the petals, thick, subulate. Although, no doubt, sufficiently hardy for the climate of the southern and western coasts of Great Britain, and also for other less favoured parts of the island when the winters are mild, we would recommend its being treated, at present, as a greenhouse plant. Experiments should, however, be made in all situations, to ascertain the degree of cold it will bear ; for if truly hardy it will prove a great acquisition to the ornamental shrubbery. Like most of the genus, it strikes freely from cuttings.—*Bot. Mag.*, t. 4626.

When we saw this in September last, in Messrs. Veitch's nursery, it was loaded with little pendulous spherical purple fruit, each having at its base the pair of bracts above described, curved back so as to resemble horns. We imagine it to be about as hardy as a common Myrtle ; but whether tender or not it is a charming acquisition, and must become a universal favourite.

482. ÆSCHYNANTH, THE SPECIES OF.—We find in the *Allgemeine Gartenzeitung* for November 22, 1851, the following list, by Mr. Edward Otto, of the Æschynanths cultivated in gardens :—

Æsch. Boschianus *de Vriese.*—Paxt. Mag. of Bot. XII. p. 176. c. tab.—Morren Ann. de la soc. d'agr. et bot. de Gand II. 403.—*Java.*
— chinensis *Gard. et Champ.* in Hort. Kew gard. misc. I. 328.—*China.*
— grandiflorus *G. Don* (Trichosporum grandifl. *Don,* olim ; Incarvillea parasitica *Roxb.* ; Æsch. parasiticus *Wall.*), Bot. Mag. t. 3843. — Bot. Reg. 1841. t. 49.—*Silhet.*
— Horsfieldii *R. Br.* (Journ. d'horticult.) Allg. Gartenz. XI. p. 243.—*Java.*
— javanicus *Hook.* Bot. Mag. t. 4503.—Van Houtte Fl. VI. 65. p. 558.—*Java.*
— Lobbianus *Hook.* Bot. Mag. t. 4261.—Van Houtte Fl. III. 246.—*Java.*
— longiflorus *Blume* (Lisionotus longifl. *Blume* olim). Bot. Mag. 4328.—Van Houtte Fl. I. c. t. 288.— Paxt. XV. 25.—*Mountain Woods of Java.*
— maculatus *Lindl.* Bot. Reg. 1841. t. 28.—*East Indies.*

Æsch. miniatus *Lindl.* Bot. Reg. 1846. t. 61.—Van Houtte Fl. I. c. t. 236. (Æsch. radicans *Wall.*—Trichosporum radicans *Blume*).—*Java.*
— pulcher *DC.* (Trichosporum pulchrum *Blum.*) Bot. Mag. 4264.— Van Houtte Fl. III. 2. t. 6.—Paxt. XVI. —*Java.*
— purpurascens *Hsskrl.* Bot. Mag. 4236. (Æsch. albida *Alph. DC.*—Bignonia albida *Blum.*—Trichosporum albidum *Nees.*—Lisionotus albidus *Blum.*)—*Java.*
— radicans *Jack.*—*Java, Sumatra.*
— ramosissimus *Wall.* (parasiticus *Hort.*) Marnock in Floricult. Mag.—*Nepal.*
— speciosus *Hook.* Bot. Mag. 4320. (Æsch. Auclandii *Hort.*) Paxt. Mag. of Bot. 1847. p. 201.—Van Houtte Fl. III. t. 267.—Ann. de la soc. d'agr. et de bot. de Gand III. 415. tab. 163.—*Java.*
— Teysmannianus *Miq.* Bot. Zeitung VI. 509. (Æsch. Teysmanni *J. Linden* Catalog. 1851.)—*Java, in woods and on the trunks of trees.*

In addition to which are the following, of which little or nothing is known :—

Æsch. atrosanguineus *Van Houtte* Cat. 1851.
— candidus *E. G. Henderson's* Cat. 1851.
— Paxtonii *Paxt.* Bot. Dict.—*Khasya.*
— pulchellus *Henders.* Cat. 1851.

Æsch. repens *Van Houtte* Cat. 1851.
— Roxburghii *Paxt.* Bot. Dict.—*Java.*
— zebrinus *Van Houtte* Cat. 1851.—*Java.*

483. PENTSTEMON BACCHARIFOLIUS. *Hooker.* A half-hardy perennial with long panicles of rich crimson flowers not unlike those of P. Hartwegii. Native of Texas. Introduced at Kew.

Stems erect, or decumbent at the base, a foot to a foot and a half high. Stem scarcely branched (except where it terminates in the panicle), terete, stout and rigid, of a purple-brown colour, and, as are the pedicels, bracts, and flowers, even the corolla within and without, clothed with minute glandular pubescence. Leaves in rather distant pairs, rigid, dark green, spreading, coarsely and spinescently toothed or serrated (generally less so at the base), glabrous : the lower ones spathulate, upwards on the stem becoming oblong, and finally, nearest the flowers, rotundate, obscurely penninerved, all of them quite sessile. Panicle terminal, elongated ; primary peduncles opposite, three-flowered, bracteated at the setting on of the peduncles and pedicels ; bracteas small, broadly ovate, reflexed. Calyx small, cup-shaped, deeply cut into five imbricating, ovate segments. Corolla rich scarlet, an inch and a half long : tube infundibuliform, labially compressed, slightly ventricose below, the mouth rather oblique, marked with a white ring : the limb obscurely two-lipped ; upper lip two-lobed, lower of three larger lobes, all patenti-reflexed. Stamens included : the fifth stamen is an abortive glabrous filament. Ovary oblong, gibbous on one side at the setting on of the long slender style : stigma capitate. This new species of Pentstemon is a native of the same region as *P. Wrightii.* Judging by the appearance of the plant after the severe frost in November last, we may conclude that it is not sufficiently hardy to live throughout the winter without some protection ; it is therefore desirable to keep a stock in pots, that may be placed in a cool frame during the winter. Being a late-flowering species, it did not ripen its seeds, but, like the allied species of the genus, it may be increased by cuttings.—*Bot. Mag.*, t. 4627.

484. DRYANDRA NOBILIS. *Lindley.* (*aliás* D. runcinata *Meisner.*) A very pretty dwarf evergreen shrub, with bright yellow flower-heads. Native of Swan River. Belongs to Proteads.

Reared from seeds sent by Mr. Drummond from the Swan River settlement. We can scarcely doubt its being the *D. nobilis* of Lindley, and of the *Plantæ Preissianæ*; yet our flowering plant, in May, 1851, was considered by Dr. Meisner (author of the *Proteaceæ* of the last-mentioned work) as a new species, which he proposed to call *Dryandra runcinata*. It is a really handsome shrub. The present plant is grown in light loam, mixed with a small portion of sharp sand. On shifting it into a larger pot or tub, we invariably keep the ball of earth an inch or more (according to the size of the plant) above the surface of the new soil; this is of importance for prolonging the life of the plant, as it prevents any excess of moisture lodging around the base of the stem. In summer, care must be taken not to allow the direct rays of the sun to strike against the sides of the pot; for the heat transmitted to the inside destroys the tender spongioles of the roots, and the plant flags and dies.—*Bot. Mag.*, t. 4633. [We confirm Sir William Hooker's determination of this plant, which appears to differ in nothing from the original specimens of *D. nobilis* now before us.]

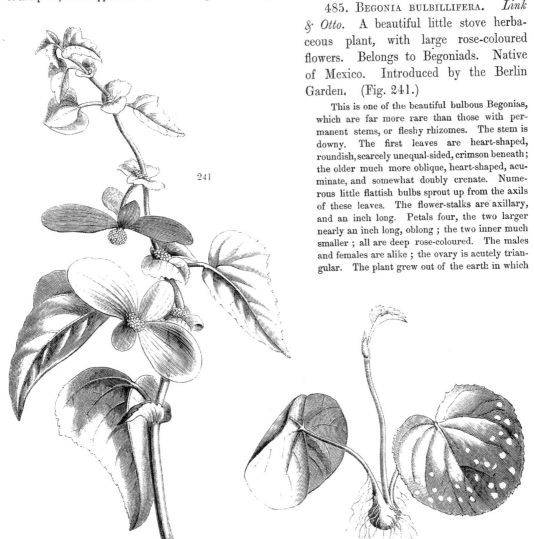

241

485. BEGONIA BULBILLIFERA. *Link & Otto*. A beautiful little stove herbaceous plant, with large rose-coloured flowers. Belongs to Begoniads. Native of Mexico. Introduced by the Berlin Garden. (Fig. 241.)

This is one of the beautiful bulbous Begonias, which are far more rare than those with permanent stems, or fleshy rhizomes. The stem is downy. The first leaves are heart-shaped, roundish, scarcely unequal-sided, crimson beneath; the older much more oblique, heart-shaped, acuminate, and somewhat doubly crenate. Numerous little flattish bulbs sprout up from the axils of these leaves. The flower-stalks are axillary, and an inch long. Petals four, the two larger nearly an inch long, oblong; the two inner much smaller; all are deep rose-coloured. The males and females are alike; the ovary is acutely triangular. The plant grew out of the earth in which some orchidaceous plant was sent from Mexico by Mr. Schiede. It flowers from August to October, in any good hot-house, or even greenhouse, provided it is planted in good rich, light earth, and has plenty of air.—*Link & Otto, Icones*, 45.

486. CASSINIA LEPTOPHYLLA. *R. Brown.* (*aliàs* Calea leptophylla *Forster.*) A hardy evergreen shrub, with clusters of white flowers. Native of New Zealand. Flowers white. Belongs to Composites. Introduced in 1824. (Fig. 242.)

We received this from an anonymous correspondent of the *Gardeners' Chronicle*. It is a dwarf, compact, heath-like bush, with dark green linear leaves, hoary, and rather yellow beneath. At the end of every branchlet is a short corymb of flower-heads, the largest of whose involucral scales are brownish, the innermost spreading and white. It is said to be a native of sandy fields near " Tolaga " and Queen Charlotte's Sound. It probably requires the same treatment as its ally, *Swammerdamia antennifera*, now beginning to be made known as a very pretty, novel, evergreen hardy bush. We see, from specimens in our possession, that the plant flowered so long ago as 1824, in the Garden of the Horticultural Society, in which it had been raised from New Zealand seeds. We also possess wild specimens from the same country from Mr. Bidwill.

242

PLATE 76.

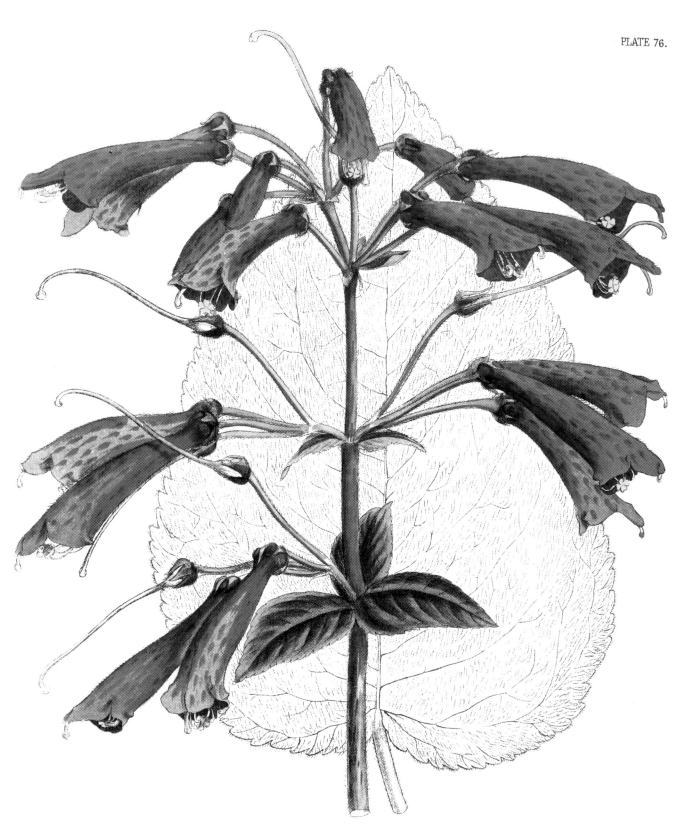

L.Constans del. & zinc.

Printed by C.F.Cheffins, London.

[PLATE 76.]

THE PURPLE GESNERA.

(GESNERA PURPUREA.)

———◆———

A noble Hothouse Tuberous Plant, of UNKNOWN ORIGIN, *belonging to* GESNERADS.

═══════════════

𝔖𝔭𝔢𝔠𝔦𝔣𝔦𝔠 ℭ𝔥𝔞𝔯𝔞𝔠𝔱𝔢𝔯.

THE PURPLE GESNERA. Leaves whorled, heart-shaped, oblong, serrato-dentate, downy. Panicle somewhat whorled, with very short peduncles. Pedicels long, umbellate, hairy. Corolla with a long tube, downy, with the upper limb straight, two-lobed, almost square, the laterals rounded and much shorter.

GESNERA *PURPUREA* ; foliis verticillatis cordatis oblongis serrato-dentatis tomentosis, paniculâ sub verticillatâ pedunculis brevibus, pedicellis elongatis umbellatis pilosis, corollis longè tubulosis tomentosis limbi laciniâ supremâ rectâ bilobâ subquadratâ lateralibus rotundatis multò brevioribus.

———————

Gesnera purpurea *of the Gardens.*

═══════════════

Tʜɪs very handsome plant belongs to the race of *G. Douglasii,* to which alone M. Decaisne limits the name, applying that of CORYTHROLOMA to *Gesnera striata, Sceptrum, ignea, Marchii,* and the like ; ISOLOMA (the Kohleria of Regel) to *G. vestita, spicata, mollis, longifolia,* &c.; DIRCÆA to *G. bulbosa, faucialis, lateritia,* &c.; while *Gesnera pardina* and *Gardneri* form the genus HOUTTEA, *G. picta* TYDÆA, *G. allagophylla* and two more RECHSTEINERA. These, and some other minor corrections necessary for restoring order among the confused mass of plants referred to Gesnera by authors, although not exhausting the subject, render the limits of the genera better than they had previously been. With the subject of the present plate, under the name of GESNERA, are associated *G. tuberosa, cochlearis, macrostachya,* and *discolor* aliàs *polyantha.*

It is evident that the present species is very near *G. Douglasii* itself, although far handsomer than even the best of the varieties (?) of that species. Not only are its dimensions larger in all

D

respects, but its flowers have a rich deep rose-colour, relieved by the characteristic spots of *G. Douglasii*, and the leaves are deeply heart-shaped, which never happens in the latter species; scarcely even in the beautiful verticillate form figured by Sir William Hooker in the *Botanical Magazine.*

But what is the history of this *G. purpurea?* It has the tender constitution and the general aspect of the tuberous stove plants with which it is associated; requiring the very same cultivation as they do. Travellers and botanists appear, however, to have been alike unacquainted with it in a wild state. Its introduction is unknown. The name which it bears seems confined to gardens, never having been registered in works of science. For these reasons we venture to suspect it to be a mere hybrid, produced perhaps between *G. Douglasii* and *G. discolor.* At all events it is one of the most striking of the noble race to which it belongs.

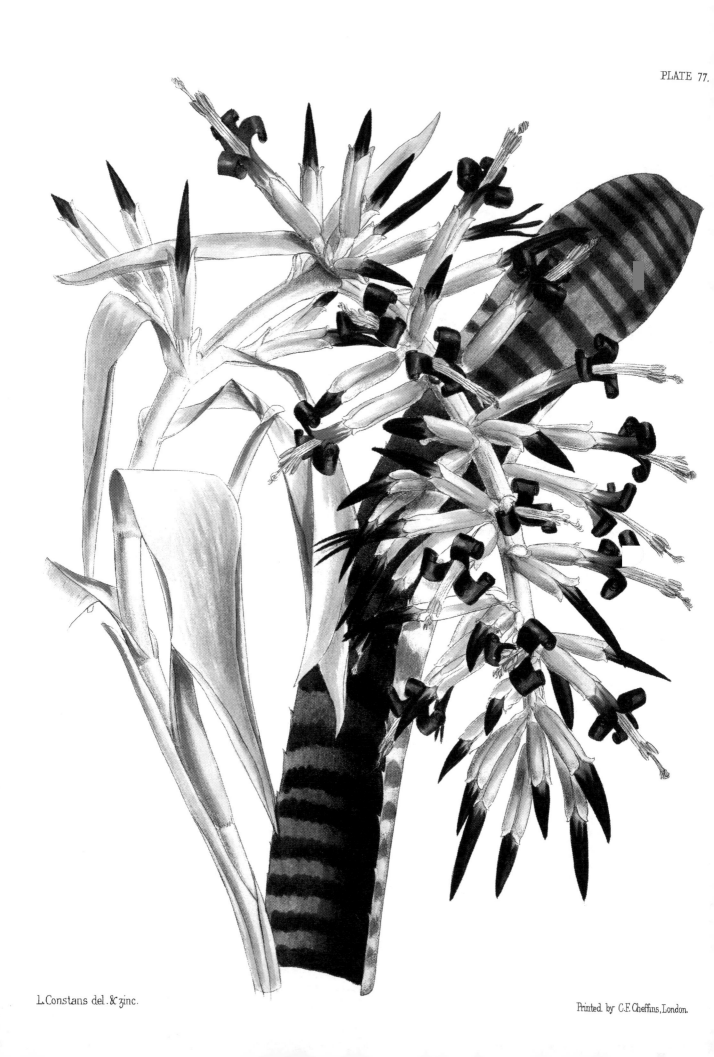

PLATE 77.

L.Constans del.&zinc.

Printed by C.F.Cheffins, London.

[PLATE 77.]

THE MOREL BILLBERGIA.

(BILLBERGIA MORELIANA.)

———◆———

A very fine Stove Perennial, from BRAZIL, *belonging to the Natural Order of* BROMELIADS.

═══════════════

Specific Character.

THE MOREL BILLBERGIA.—Leaves strap-shaped, channelled, blunt, banded with white, as long as the stem, with some spiny teeth near the base. Stem smooth, clothed with large loose petaloid distant scales. Raceme many-flowered, recurved, nearly smooth. Bracts coloured, finely scaly at the back, longer than the fascicled flowers. Sepals oblong, obtuse, mucronate, with a membranous margin, smooth, as well as the ovary. Petals revolute, much longer than the sepals. Stamens projecting far.

BILLBERGIA *MORELIANA ;* foliis ligulatis canaliculatis obtusis albo-fasciatis versus basin spinoso-dentatis cauli æqualibus, caule glabro squamis magnis petaloideis laxis distanter vestito, racemo multifloro recurvo glabriusculo, bracteis coloratis dorso minutissimè lepidotis floribus fasciculatis longioribus, sepalis oblongis obtusis mucronatis membranaceo-marginatis ovarioque lævibus, petalis revolutis calyce multò longioribus, staminibus longè exsertis.

───────────

Billbergia Moreliana : *Adolphe Brongniart* in *" Portefeuille des Horticulteurs."* *Revue Horticole*, iii., 82.

═══════════════

ONE of the most charming of the Bromeliaceous Order, and among the easiest to cultivate. Its flaming rose-coloured bracts contrast finely with the deep clear violet of the petals, and appearing on drooping racemes above a foot long, produce an unusual as well as most brilliant effect.

The species appears to be a native of Brazil. It was originally published by Prof. Adolphe Brongniart in the *Portefeuille des Horticulteurs,* a work we have not seen. Shortly afterwards it was mentioned in the *Revue Horticole* in the following terms : —

" This magnificent Bromeliad is cultivated by M. Morel, a zealous amateur, possessing the most beautiful collection of Epiphytes in Paris. In its leaves, the species which we describe reminds us of certain Tillandsias destitute of spiny teeth ; but the flower-stem, turned back, branching, and furnished at the upper end with large bright rose-coloured delicate and semi-transparent bracts, covered with a white mealy powder, immediately distinguishes it.   From the axil of these bracts spring the flowers, which are slightly irregular, of a pure violet colour, rendering this species one of the most beautiful ornamental plants of our hot-houses.   M. Morel cultivates it in baskets, hung up, and filled with peat earth covered with Lycopodium, which retains the freshness of the soil, and at the same time indicates the moisture of the house."

We find no other notice of the plant.   The specimen now represented was flowered in the garden of the Horticultural Society, where it had been received from M. Keteler, of Paris, in 1848, as a fine variety of *Billbergia zebrina*.   In February last we observed it in flower with Messrs. E. G. Henderson and Co., of the Wellington Nursery, St. John's Wood, who obtained it from M. Morel himself.

As to *Billbergia zebrina*, of which it has been supposed to be a variety, it is enough to observe that the ovaries and sepals of that plant are closely coated with white meal, and the stamens twice as long as in the plant before us, to say nothing of the leaves of *Billbergia zebrina* being spiny to their points, and the bracts by no means so richly tinted.

PLATE 78

L.Constans del. & zinc.

Printed by C.F.Cheffins, London.

[PLATE 78.]

THE MASTERS CYMBID.

(CYMBIDIUM MASTERSII.)

———◆———

A handsome Terrestrial Orchid, from the EAST INDIES.

Specific Character.

THE MASTERS CYMBID. Leaves distichous, narrowly sword-shaped, obtuse. Peduncle erect, closely covered with herbaceous equitant sharp-pointed scales. Spike short, few-flowered, plunged within the scales. Sepals and petals linear-oblong, blunt. Lip obovate, three-lobed, downy inside; with the ridges continuous, confluent at the points, and sometimes expanded into a three-lobed tubercle; the middle segment oblong, wavy, lobed, those at the side blunt and flat.

CYMBIDIUM *MASTERSII*; foliis distichis angustè ensiformibus obtusis, pedunculo erecto squamis herbaceis equitantibus acutissimis imbricato, spicâ brevi paucifiorâ squamis immersâ, sepalis petalisque lineari-oblongis acutis, labello obovato trilobo intùs pubescente, lamellis continuis apice confluentibus nunc in tuberculum subtridentatum expansis, laciniâ intermediâ oblongâ undulatâ lobatâ lateralibus obtusis planis.

Cymbidium Mastersii : *Griffith in Hort. Bot. Calcutta; Loddiges' Catalogue,* No. 1233; *Lindley in Botanical Register,* 1845, t. 50.

WHEN this was published in the Botanical Register, seven years ago, nothing could be said about it except that it was received from the East Indies by Messrs. Loddiges in the year 1841, and blossomed in December, 1844; that it is a very distinct species, with snow-white flowers, sweet-scented, having the fragrance of almonds; and that its erect flower-stalk, closely covered with long green sharp-pointed equitant imbricated sheaths, is quite unlike that of any other species. It was understood to have been named by Griffith after Mr. Masters, one of the principal assistants in the Botanical Garden, Calcutta.

Since that period it has continued to appear occasionally in collections, but remains a rare plant. The specimen now figured, if compared with the original plate in the Botanical Register, will show what cultivation has done in the hands of Mr. Bateman, from whom we received it last December.

It is undoubtedly a genuine Cymbidium, as is shown by the two parallel plates on its lip, and the short somewhat transverse gland of the pollen masses. One of its nearest affinities is *C. elegans*, another species from the continent of India.

Although the species of this genus are capable of growing upon the bark of trees, and the Aloe-leaved was one of the very few which was able to endure the ill-treatment of gardeners before 1822, yet they are much more advantageously regarded as terrestrial plants. They should all be grown in pots, in thoroughly-drained lumps of peat, into which their long roots can penetrate, roasted in summer, but well watered and kept in an atmosphere saturated with humidity, but continually in motion while they are making their growth, after which they should be gradually dried off again.

GLEANINGS AND ORIGINAL MEMORANDA.

487. CHEIROSTEMON PLATANOIDES. *Humboldt & Bonpland.* A large greenhouse tree, with broad Plane-like leaves, and brown bell-shaped flowers. Native of Guatemala. Belongs to the Order of Sterculiads. (Fig. 243.)

In a recent number of the *Flore des Serres* are some observations on this plant by M. Adrien de Jussieu, from which we borrow our figure and what follows. The tree, known among us as "the Hand-plant," has never flowered in our collections, although common in them.

This tree was observed in Mexico from the time when that country was discovered. The natives made a kind of pilgrimage to it, and collected the flowers for amulets. The more interest attached to it on account of its rarity. According to Humboldt (*Tableaux de la Nature*, 2, p. 161), " but one solitary individual existed in the whole Mexican Confederation,—one ancient stock of this marvellous plant. It was believed that the tree had been planted 500 years before by a king of Toluca, as a specimen of exotic vegetation. But how was it that one individual only was known? and whence came the young plant, or its seed? It was difficult to understand why Montezuma had it not in those Botanical Gardens of Huaxtepec, Chapoltepec, and Iztapalapan, of which Hernandez made such good use, and of which some traces still remain. — It is said to be wild in the forests of Guatemala." [Since Humboldt wrote, the accuracy of his conjecture has been established. Hartweg found it on the mountains of Acatenango, and on the volcano called the Volcan de Agua, forming trees from fifty to eighty feet high.] Hernandez made the plant known in his celebrated work (*Rerum medicarum Novæ Hispaniæ Thesaurus*) by a short description and figure. He preserved the Mexican name Macpalxochiquahuit, a name having the same meaning as the Arbol de las Manitas of the Spaniards, or Hand-plant, so called on account of the five stamens being joined together, and, on their emerging from the dull purple calyx resembling a hand, or rather a paw with five claws. In the garden of Montpellier is a tree raised from seeds obtained from Madrid ; in 1813 it was twenty feet high, but had not flowered ; but since that time it has blossomed abundantly. At Paris it has of late years been planted out in one of the great conservatories, is now about thirty feet high, and has occasionally flowered since June, 1850. To this M. de Jussieu adds a detailed botanical description, for which the reader is referred to the *Flore des Serres* itself, vol. vii., p. 8, &c.

243

488. Passiflora sicyoides. *Schlechtendahl.* (*aliàs* P. odora *Link & Otto.*) A greenhouse twining plant, with greenish sweet-scented flowers. Native of Mexico. Introduced by the late George Barker, Esq., of Birmingham. Flowers in August. (Fig. 244.)

This fragrant climber first flowered with Mr. Barker in 1839, since which time it has continued to appear occasionally, although we do not find it figured in any English work. The whole surface is covered with forked hairs. The leaves are heart-shaped, three-lobed, with the middle lobe the longest and entire ; the lateral lobes, which are placed at right angles with it, are pretty generally furnished with bristle-pointed serratures near the base. Their stalk is remarkable for two large opposite, oblong, glands. The flowers are solitary in the axil of the leaves, on stalks shorter than the petioles ; with four very long hairy deciduous bracts. The sepals are greenish, hairy outside, white within ; the petals are much smaller and white. The coronet consists of threads, variegated with red. The fruit is bluish black, about as large as an Orleans plum, and readily separates into valves. The seeds are roundish, pale gray, with very deep pits. Messrs. Schiede and Deppe found the plant near Jalapa. It has been distributed among Coulter's Mexican plants under two numbers— 62 and 63. Why Messrs. Link and Otto altered Professor Schlechtendahl's name, we are unable to explain.

244

489. Ranunculus cortusæfolius. *Willdenow.* (*aliàs* R. Teneriffæ *Pers.* ; *aliàs* R. grandifolius *Low.*) A large-flowered hardy perennial, with a weedy habit. Native of the Canary Islands, &c. Blossoms yellow.

Unquestionably the handsomest of all the Buttercups yet known to botanists. The flowers are not only large, more than two inches across, but of a singularly glossy yellow colour ; and although a native, as it would seem exclusively, of the Canary Islands and of Madeira, it is quite hardy. In the latter country Ribeira Frio seems to be the only locality : in the former, Mr. Webb describes it as inhabiting grassy banks in the woody districts. It flowers during the summer months. This plant, being of neat habit and flowering freely in a pot, is well suited for being associated with general collections of the smaller alpine plants, which are usually kept in pots for the convenience of removing the more tender species to the protection of a frame during the winter and early spring months. When planted in the open border, it should be protected by a hand-glass, additional covering being provided during severe frosts. It is increased by division of the roots, which should be done in autumn.—*Bot. Mag.*, t. 4625.

490. Viola pyrolæfolia. *Poiret.* (*aliàs* V. maculata *Cavanilles* ; *aliàs* V. lutea *of Gardens.*)

A very handsome stemless hardy perennial plant, with large yellow flowers. Native of Patagonia. Introduced by Messrs. Veitch.

Some of our readers have seen this beautiful species in the exhibitions of London. It is a stemless hairy herbaceous plant with leaves not unlike those of the dog-violet in form, but thicker, convex, and a good deal wrinkled. From them rise up numerous stout flower-stalks, each three or four inches high, and bearing a single large pure yellow firm blossom, with a short blunt spur. The reason which led to the name *maculata*, or spotted, given this plant by Cavanilles, is thus explained by Dr. Planchon in the *Flore des Serres*, where it is admirably figured :— " Various parts of the tissue of violets, especially the parenchym of the leaves, sepals and seed-vessels, contain small heaps of what seems to be resinous matter, which, in dried specimens, manifests itself on the lower surface of the leaves, the sepals, and the capsules in the form of minute brownish points. These vary in form from the round point, to short lines, such as occur on Anagallis and Parnassia ; they are sprinkled over most violets, and though scarcely apparent in the generality of species, become numerous and very evident in the plant before us, but only on the old and dried leaves. M. van Houtte states that the *Viola lutea* does very well out of doors if treated like an alpine plant, that is to say, kept in a cool shady place, in light soil consisting chiefly of black vegetable mould, and well drained. If grown in pots it flowers freely in the greenhouse during winter. It is propagated by side runners."

491. DENDROBIUM BIGIBBUM. A tropical New Holland epiphyte, with pretty purple flowers. Introduced by Mr. Loddiges. Blossoms in January. (Fig. 245.)

D. bigibbum (Dendrocoryne) ; caulibus elongatis apice 3-5-phyllis, racemis erectis elongatis dissitifloris, petalis subrotundis sepalis duplò latioribus, labelli trilobi lobis rotundatis medio cristato basi gibboso, sepalis lateralibus in calcar productis.

This very remarkable plant was obtained from the north-west coast of New Holland by Mr. Loddiges, with whom it flowered in January last. The stems are long, narrow, fusiform, or tapering to the base, closely invested with dry light brown sheaths ; near the end they bear five or six long, narrow, firm, spreading acute leaves, each with *five* ribs (not three as in the accompanying cut). The raceme is erect, and consists of three or four flowers, placed at the end of a graceful peduncle eight or nine inches long. The bracts are small and scale-like. The blossoms are rich purple, nearly of the same colour as *Bletia verecunda*. The sepals are oblong, acute, flat ; the lateral ones united at the base, so as to form a short blunt spur below the setting on of the labellum. The petals are roundish, and slightly recurved. The lip originates in the sinus above the spur of the sepals, is moveable, and projects outwards at its base in the usual way, so that this flower has a kind of double chin. The three lobes of the lip are rounded and of nearly equal size, the central one being the darkest colour ; along the middle are three raised lines, which terminate at the base of the central lobe in the form of three short rows of fleshy notches. At the base of the lateral sepals next the orifice of their spur is found on each side a thick callus. The species is nearly allied to *D. Kingianum* and *elongatum*, but is much handsomer.

492. ROSCOEA PURPUREA. *Smith.* A half-hardy herbaceous plant with dark purple flowers. Native of Khasya. Belongs to Gingerworts. Blossoms in September.

Reared from tubers sent to the Royal Gardens, from Khasya, in North-eastern Bengal, by Dr. Hooker ; and these specimens (flowering in September, 1851) exactly correspond with drawings made by that naturalist on the spot. They sufficiently accord with the original *R. purpurea* of Sir J. E. Smith, to satisfy us that it is identical with that species ; whereas, further north, in Sikkim-Himalaya, Dr. Hooker detected and drew and transmitted living plants to Kew of what has been called *R. purpurea* by us (in *Exotic Botany*), by Mr. Roscoe (in his fine work on Monandrian plants), and by Dr. Lindley (in *Botanical Miscellany*). All the plants of these authors agree in being larger and stouter than the one now before us, with swollen stems and ovato-lanceolate approximate leaves, and flowers of a pale lilac-purple, with a very large and broad lip, nearly entire at the apex. We hardly dare venture to assert that the two kinds are truly distinct, though I am disposed to think them so : but whether species or varieties, our present plant, now we believe first reared in England, is the same with the original *R. purpurea*. This Himalayan species is sufficiently hardy to thrive in a cool pit, protected from frost. After the decay of the stems, the underground tuber-like rhizome remains in a dormant state during the winter. At this season the soil in the pots should be kept just sufficiently moist to preserve the tubers from shrivelling. Early in the spring these should be repotted in fresh soil, consisting of a mixture of light loam and peat, little or no water being given till they begin to grow, and then but sparingly ; for being of a soft fleshy nature, the tubers are liable to rot off through any excess of moisture.—*Bot. Mag.*, t. 4630.

493. CATALPA POTTSII. *Seemann.* A half-hardy shrub, from Mexico. Belongs to Bignoniads. Flowers apparently pink. Introduced at Kew.

A bush four to six feet high. Branches very smooth. Leaves coriaceous, linear-lanceolate, entire, glaucous. Flowers from two to two and a half inches long.

Two species of Catalpa, viz., *C. syringæfolia* Sims, from North America, and *C. longissima* Sims, from the West Indies, have been for some time cultivated in the gardens of Europe. To these has been lately added a third from Chihuahua, one of the northern states of Mexico. It was raised by Mr. F. Scheer from seed, sent over in 1850 by Mr. John Potts, and is now to be found in the gardens at Düsseldorf, Hanover, and Leipsig. This circumstance has induced me to name it, and I have accordingly done so after its discoverer. To the above short character, a more detailed account will be added in " The Botany of H. M. S. Herald."—*Seemann, in Allgem. Gartenzeit.,* Oct. 11, 1851.

494. RYTIDOPHYLLUM OERSTEDTII. *Klotzsch.* A half-shrubby hothouse plant, with greenish flowers spotted with purple. Belongs to Gesnerads. Native of Central America. Introduced by M. Warczewicz.

Rytidophyllum Oerstedtii ; suffruticosum ; caule erecto, ramoso, hirto ; foliis oblique oblongis, membranaceis, inter se inæqualibus, simpliciter serratis, longissime acuminatis, basi attenuatis, supra sparsim pilosis, subtus nervoso-villosis ; petiolis distinctis, villosis ; corymbis in apice ramulorum axillaribus, longe pedunculatis, folio subduplo longioribus, 3—5 floris, subaphyllis ; calycis laciniis lanceolatis, acutis, trinerviis, utrinque germineque villosis, tubo corollæ magis hirsuto, duplo brevioribus ; corollæ lobis subglabris, rotundatis, virescentibus, purpureo-maculatis ; filamentis superne pilosis ; stylo sparsim et brevissime piloso ; stigmate incrassato ; seminibus minutissimis, scobiformibus, in capsulam uniloculorem obpyramidalem apice bivalvem parietalibus ; disco epigyno annulari, 5-lobo.

A half-shrubby plant, two feet high, throwing out, like an epiphyte, a quantity of air-roots, which stick close to the bark of the tree on which the plant stands ; for this reason, and because it grows at a considerable height from the ground, and is thus subject to no inconsiderable daily change of temperature, its cultivation is difficult. The plant was discovered by Dr. Oersted and M. von Warczewicz, in Costa Rica (Central America) growing on trees. Living specimens were first introduced by M. von Warczewicz ; and M. Nietner, of Schönhausen, near Berlin, has now plants in flower, raised from seeds obtained from him. The leaves are from three to seven inches long, and from one to two and a half inches broad. The petiole is from half an inch to one inch long. The hairs are, as in all the species belonging to this genus, jointed. The flowers are an inch and a half in length, the tube swollen and bent, three quarters of an inch in diameter. The corolla is hairy, with broad, rounded, distinct lobes, green, with purple spots.—*Klotzsch, in Allgem. Gartenzeit.,* Jan. 17, 1852.

495. LENNEA ROBINIOIDES. *Link, Klotzsch & Otto.* A Mexican greenhouse tree, with the

appearance of a Robinia. Flowers purple. Belongs to the Leguminous Order. Introduced by the Royal Garden, Berlin. (Fig. 246.)

This is a small tree, in cultivation a mere bush, from two to three feet high, destitute of hairiness, with unequally pinnated distichous leaves. Stipules free, subulate, deciduous. Leaflets in four or five pairs, with prickly stipules at their base. Racemes axillary, pendulous. Flowers as large as those of the Judas tree, and of the same colour, appearing in May. The genus is recognised by Mr. Bentham, who places it between Robinia and Sabinea. It lives out of doors at Berlin in the summer, although requiring there the shelter of a greenhouse in the winter.

496. ODONTOGLOSSUM EHREN-BERGII. *Klotzsch.* A Mexican Orchidaceous epiphyte, with delicate white flowers spotted with brown on the sepals. Introduced in 1846. Flowers in August. (Fig. 247.)

C. Ehrenbergii (Leucoglossum); pseudobulbis caespitosis, globoso-subelongatis, compressis ; foliis solitariis, ellipticis, acutis, membranaceis, rigidis, margine subreflexis ; scapo unifloro, medio articulato, bibracteato ; perigonii foliolis candidis, exterioribus lanceolatis, acuminatis, dorso longitudinaliter carinatis, patentibus ; interioribus latioribus, oblongis, acutis, utrinque attenuatis, recurvis ; labello subcordato, acuto, undulato, crenulato; lamellis unguis callosis, integerrimis, antice in rostrum obtusum breve confluentibus ; columna aptera, puberula.—*Klotzsch abbrev.*

This is one of the prettiest of the white-lipped Odontoglots. M. Charles Ehrenberg found it on an oak tree near San Onofro, on the banks of the River Zimapore. In habit it is hardly distinguishable from *O. Rossii*, to which we formerly referred it ; but it seems to have a dwarfer habit, smaller flowers, and especially thin delicate white sepals banded with brown, instead of green ones ; the lip too is acuminate, not rounded, each stem bears but one flower, and the processes at the base of the lip are white, not yellow, and join into an undivided apex instead of a two-lobed one. Perhaps as good a way of bringing this species distinctly to the reader's eye is to speak of it as being intermediate between *O. Rossii* and *O. stellatum*.

497. MACHÆRANTHERA TANACETIFOLIA. *Nees.* (*aliàs* Aster tanacetifolius *H. B. K.*; *aliàs* A. chrysanthemoides *Willldenow.*) A handsome half-hardy suffruticose plant, with large deep-violet flower-heads. Belongs to Composites. Native of New Mexico. Introduced at Kew.

A pretty and singular suffruticose Composite, with flowers nearly as large as a China Aster, and the leaves deeply pinnatifid, like some Anthemis, perhaps, rather than Tanacetum. It was seen by Humboldt cultivated in gardens in Mexico ; but Dr. Wright appears to have found it wild in Mew Mexico, and from his seeds our plants were raised in the Royal Gardens of Kew. Planted in the open border they continued flowering during the summer months. A procumbent, or rather ascending, half-shrubby plant, with branching slender stems, nearly a foot long, everywhere, as well as the foliage, slightly downy. Flower-head large, yellow, with a purple ray, solitary, terminal on the branches. Involucre hemispherical, of numerous, spreading, subulate, glandular, herbaceous, scales. Ligules of the ray rather linear-lanceolate, three-nerved, the lower portion woolly at the back. Achenia hispid. This pretty plant is a tender biennial, but sufficiently hardy to flourish in the open air during summer. Unfortunately for its maintenance as a garden plant, it produces but a small quantity of perfect seeds, and is not readily propagated by cuttings.—*Bot. Mag.,* t. 4634. [Otherwise it would be a fine bedding out plant, its colour being one much wanted in gardens.]

498. TRICHOPILIA ALBIDA. *Wendland.* A stove epiphyte, with white and yellow flowers. Belongs to Orchids. Native of the Caraccas. Introduced by M. Otto, of Hamburgh.

T. pseudobulbis oblongo-lanceolatis, compressis, sulcatis, monophyllis; foliis oblongo-lanceolatis, planis, basi subcordatis, apice acuminatis, recurvis ; racemis basilaribus pendulis, subtrifloris ; perigonii foliolis conformibus, lineari-lanceolatis, acuminatis, undulatis, rectiusculis, subtortis, pallide luteo-viridulis, margine subhyalinis ; labello petalis longiore, quadrilobo, lobis rotundatis undulato-crispatulis, basi arcte convoluto, albido, fauce punctis luteo-ochraceis confluentibus adspersâ ; cucullo trilobo, laciniis fimbriatis, mediâ longiore.

The compressed pseudobulbs are five inches long, and from six to ten lines broad, flat, somewhat furrowed and sharp-cornered, oblong and a little narrow towards the top. The young inflorescence is covered by darkly-spotted sheaths. The leaves are a little longer than the bulbs, from an inch to an inch and a half broad, solitary, leathery, somewhat heart-shaped and downy at their base, flat, and with recurved points. The flower-spikes, which generally bear three flowers, proceed from the base of the pseudobulb, are from four to six inches long, and of the thickness of a crow-quill. The flower is three inches in diameter ; the sepals and petals are alike, an inch and a half long, and three lines broad, linear-lanceolate, pointed, waved at the edge, tolerably erect, but inclined a little forwards, not much twisted, pale yellow-green, and nearly transparent at the edge. The labellum is smooth, a little longer than the sepals, four-lobed ; the lobes are rounded, waved, and crumpled at the edge, and rolled closely together at the base ; in the middle of the labellum are a few irregular raised longitudinal streaks. The colour of the flowers is white, with a large spot in the middle, made up of a quantity of small, crowded yellow ochre-coloured points. The column is straight, white at the top and light green towards the base. The hood is three-lobed, the middle lobe being a little prominent, and all fringed. The flowers have a faint delicate odour, and last only a few days. This species is closely allied to *Trichopilia tortilis* Lindl. and *T. coccinea* Lindl., and is distinguished from them, independently of the colour of the flowers, by its longer pseudobulbs, and by its scarcely twisted petals. It was imported in July,1851, with other Orchids, from M. Wagener in the Caraccas, consigned to M. C. Otto of the Hamburgh Botanic Garden.—*Wendland, in Allgem. Gartenzeit., Nov.* 15, 1851.

499. CANNA SANGUINEA. *Warczewicz.*

Concerning this, which is not the *Canna sanguinea* of others, we find the following memorandum in the *Allgemeine Gartenzeitung* for September 13th, 1851 :—

"This new species, from Costa Rica, was introduced into the gardens of Germany by M. Warczewicz in 1850. It is one of those which succeed in the open ground in summer ; it flowers freely, and is remarkable for its beautiful blood-red blossoms. In autumn it should be taken up and kept all the winter in a temperate greenhouse. If divided and forced in March or April, an early flowering may be expected. To be seen in all its beauty, the plant requires a warm sheltered place, rich garden mould, and a plentiful supply of water. It seeds abundantly. The specimens which we saw in M. Mathieu's garden were three feet high."

500. CYCNOCHES MUSCIFERUM. A curious epiphyte from Colombia, with pale flowers spotted with brown. Flowers in February. Introduced by Messrs. Rollissons from Mr. Linden. (Fig. 248.)

C. musciferum ; racemo laxo stricto, bracteis subulatis, sepalis lineari-lanceolatis acutis dorsali refracto, petalis linearibus, labello membranaceo hastato : laciniis lateralibus linearibus ascendentibus intermediâ basi rhombeâ barbatâ in apicem linguiformem attenuatâ.

    This very curious little plant looks like a diminutive form of *C. barbatum ;* its flowers are very pale bistre plentifully bestrewed with minute brown specks and freckles. It is a curiosity, but not brilliant enough in appearance to suit the taste of any except botanists. The resemblance of the blossoms to some kind of fly is striking.

    501. SISYRINCHIUM MAJALE. *Link, Klotzsch & Otto.* A half-hardy perennial, from Chili, belonging to the Order of Irids. Flowers yellow with a brown eye. (Fig. 249.)

A dwarf perennial, with rough narrow grassy leaves, and large rough green spathes, from among which the flowers appear in succession for some weeks in May and April. The roots are fleshy and fasciculate ; the stem is from six inches to one foot and a half high ; the sepals and petals bright yellow, with a deep brown spot, variable in size, at the base of each. This is no doubt the *Sisyrinchium graminifolium,* var. *pumilum,* of the *Botanical Register,* t. 1914 (1915), of which specimens are before us from Conception, where they were gathered by Macrae. The true *S. graminifolium* is represented by No. 478 of Cuming's Chilian Collections. According to Dr. Klotzsch, the species most nearly allied to *Sisyrinchium majale* are *S. tenuifolium, convolutum, palmifolium, flexuosum* and *graminifolium. S. tenuifolium* is distinguished by its entire-edged calyx ; *S. convolutum*, by its fibrous roots ; *S. palmifolium*, by its bulbous root and white flowers ; *S. flexuosum*, by its crooked stem and its densely hairy ovaries ; and *S. graminifolium*, with its varieties, only by its undivided bracts, cylindrical smooth stem, and rough leaves.

502. PENTARHAPHIA VERRUCOSA. *Decaisne.* (*aliàs* Conradia verrucosa *Scheidweiler.*)   A rigid greenhouse shrub, with tubular scarlet flowers.   Belongs to Gesnerads.   Native of Cuba.   Introduced by Linden.   (Fig. 250.)

There is a plant not uncommon in gardens, called *Pentarhaphia cubensis,* which is so like this as to suggest the possibility of the two belonging to the same species.   Both were found by Mr. Linden in Cuba : that now figured, on Mount Liban, flowering in May ; the other at a place called Pinal de Nimanima, both in the province of St. Jago.   They differ, however, in the branches of *P. verrucosa* being covered with little tubercles and being much blistered (bullate) in consequence of the parenchym of the interspaces growing faster than the veins ; while in *P. cubensis* the branches are smooth, the leaves flat, and the flowers larger.   Both are useful hard-leaved greenhouse shrubs, quite different in constitution from the soft-wooded species of the same natural Order.   The genus to which they are referred received its name twenty-five years since, from the writer of the present note, the *Gesnera ventricosa* of Swartz, upon which it was founded, being the only species known to him.   At a later period Prof. von Martius proposed to abolish *Pentarhaphia*, and to create a genus *Conradia,* in which it was to merge.   Notwithstanding the obvious objections to this measure, De Candolle unadvisedly acquiesced in it.   But Prof. Decaisne, in a luminous paper in the *Annales des Sciences* for 1846, restored the genus *Pentarhaphia*, increasing the number of its species to fifteen, and left Von Martius' name of Conradia for one species only, the *Gesnera humilis* of Linnæus.   The genus *Pentarhaphia* still then continues to be known by the five long needle-like teeth of its wholly inferior calyx, its five to ten-ribbed fruit, and its annular disk. The wild specimens of *Pentarhaphia verrucosa* brought from Cuba by Mr. Linden, are covered with a glutinous exudation, and the leaves are much harder, stiffer, and more bullate than in the garden plant.

250

PLATE 79.

L.Constans del.& zinc.

Printed by C.E.Cheffins, London.

[PLATE 79.]

THE NEPAL ASH-LEAVED BERBERRY.

(BERBERIS NEPALENSIS.)

◆

A half-hardy Evergreen Shrub, with yellow flowers, belonging to BERBERIDS, *from the* EAST INDIES.

Specific Character.

THE NEPAL HOLLY-LEAVED BERBERRY. Leaves pinnated, leaflets in from two to five pairs, ovate, spiny-toothed, with the odd one on a long stalklet. Racemes fascicled, upright, compactly flowered. Fruit oblong.

BERBERIS *NEPALENSIS;* foliis pinnatis, foliolis 2—5-jugis ovatis spinoso-dentatis cum impari petiolulato, racemis fasciculatis strictis densifloris terminalibus, fructu oblongo.

Berberis pinnata : *Roxb. fl. Indica*, ii. 184. Mahonia nepalensis : *De Cand. Prodr.* i. 109. Berberis nepalensis : *Wallich Catalogue, no.* 1480 ; *Lindley in Hort. Soc. Journal*, vol. v., p. 18.

THIS beautiful specimen of one of the handsomest of the pinnated Berberries was produced in the Garden of the Horticultural Society in March last. It had been received from the Royal Botanic Garden, Kew. About the same time it blossomed in several other places, we believe for the first time in Europe. It is very near the North American *B. glumacea.*

When grown in a conservatory the species is remarkable for the delicate light green of its foliage, which spreads gracefully from a stiff erect stem, something in the way of a miniature Palm. At first the plant produces its leaflets in threes; at a late period they grow in fives, and when in complete vigour they appear in about five pairs with an odd one. Each leaflet is very regularly furnished with large equal spiny teeth along the whole of its ovate or ovate-oblong outline. The flowers are of a rich bright yellow, forming close erect racemes clustered in the upper end of the shoots, and drooping gracefully. Their ovary is oblong.

F

The plant is probably hardy; at least it has sustained no injury during one winter in the open air, and a slight screen of glass without fire has saved it from the effects of the unprecedented cold of the present spring. But it is doubtful whether it will not be necessary to give it the protection of a glass roof, in order that its beautiful leaves may not be injured by winds. It is understood to prefer sheltered nooks in the Himalayas, and there only to display the beauty that belongs to it.

It seems probable that Asia contains four Berberries nearly related to this, if not five, all of which would prove horticultural treasures.

First, there is the present plant, which seems to be confined to the chain of the Himalayas and the adjoining districts.

A second is the *B. acanthifolia* of Wallich, abundant in the Nilgherry range; when growing in favourable situations, as Dr. Wight informs us, it forms a small tree. It is known by its very numerous leaflets, as many as twenty-one in some specimens, and bluish-purple globose, not oblong, fruit. It appears to be the same as *B. Leschenaultii* of Wallich and Wight, which the latter finds in almost every clump of jungle about Ootacamund, the well-known sanitarium of the Madras presidency.

A possible third is mentioned by Dr. Wight as having drooping racemes, and inhabiting Coorg. He supposes it to be identical with a plant seen by him on the Pulney Mountains, with " diffuse rambling branches."

A fourth is the *Berberis japonica*, figured at No. 10 of the Gleanings in our first volume.

A fifth is a most remarkable species, found by Mr. Fortune in his visit to the tea countries of China, and regarded by him as a possible form of *B. nepalensis*. Of this *B. trifurca* we shall speedily produce a figure.

PLATE 80.

L.Constans del.& zinc.

Printed by C.F.Cheffins,London.

[PLATE 80.]

THE MANY-SPIKED BILLBERGIA.

(BILLBERGIA? POLYSTACHYA.)

◆

A handsome evergreen Hothouse Perennial, Belonging to BROMELIADS, *from* BRAZIL.

Specific Character.

THE MANY-SPIKED BILLBERGIA. Leaves channelled, with spiny teeth, curved back at the point, inflated at the base, shorter than the scape. Spike conical, many-ranked, mealy. Bracts roundish, acuminate, closely imbricated.

BILLBERGIA ? *POLYSTACHYA ;* foliis canaliculatis spinoso-dentatis apice recurvis basi ventricosis scapo brevioribus, spicâ conicâ polystachyâ farinosâ, bracteis subrotundis acuminatis arctè imbricatis.

Our knowledge of this beautiful plant is very imperfect. A specimen in flower was exhibited by M. de Jonghe, of Brussels, at one of the Meetings last year in the Garden of the Horticultural Society, as a new species of Billbergia. Having been afterwards removed we had no opportunity of describing it, and are only now able to make it known by means of a coloured drawing which accompanied the specimen.

It is no doubt a Brazilian plant, and seems nearly related to Lemaire's *Billbergia rhodocyanea,* another charming species, figured in the *Flore des Serres,* vol. iii., p. 207, with long loose stiff spiny-toothed crimson bracts, bright blue corollas, and broad blunt dark green leaves banded with white. That plant flowered with Mr. Van Houtte, but has not appeared in our gardens.

PLATE 81.

L.Constans del.& zinc.

Printed by C.F.Cheffins,London.

[PLATE 81.]

THE ROSY LIMATODE.

(LIMATODES ROSEA.)

———————◆———————

A most beautiful Terrestrial Hothouse ORCHID *from the* EAST INDIES.

═══════════════════

Specific Character.

THE ROSY LIMATODE. Pseudobulbs fusiform. Leaves oblong-lanceolate, plaited, smooth. Scape many-flowered, longer than the leaves, shaggy, as well as the loosely placed flowers. Bracts membranous, curved backwards, shorter than the ovary. Lip oblong, flat, retuse. Spur straight, blunt, horizontal. Column dwarf, downy.

LIMATODES *ROSEA ;* pseudobulbis fusiformibus, foliis oblongo-lanceolatis plicatis glabris, scapo multifloro foliis longiore floribusque laxis villosis, bracteis membranaceis recurvis ovario brevioribus, labello oblongo plano retuso, calcare recto obtuso horizontali, columnâ nanâ tomentosâ.

═══════════════════

THE genus Limatodes has hitherto been known to the public exclusively by a figure in the plates belonging to Blume's *Bijdragen,* and the scanty accompanying letter-press. The species there mentioned, *L. pauciflora,* a native of dense woods on Mount Salak in Java, is described as a fibrous-rooted terrestrial plant, having stems swollen at the base, broadly lanceolate membranous ribbed leaves, lateral solitary few-flowered peduncles (by which we understand scapes), and white blossoms. The figure shews it to be a genus very nearly allied to Calanthe, from which it differs in having the lip perfectly free from the column, instead of being united with it. It also appears to have a column much elongated, while that of Calanthe is in general particularly short ; but such a difference is unimportant, because *Calanthe densiflora* has also a very long column, and the discovery of the present species with a very short column still further destroys any value which the character alluded to may have been supposed to possess.

It was near Moulmein, in the province of Martaban, that this brilliant species was discovered by Mr. Thomas Lobb, and sent to Messrs. Veitch, with whom it flowered in December last. In all respects it has the habit of a Calanthe, but the pseudobulbs are long and fusiform. The stem and flowers are covered with long hairs like *Calanthe vestita*. The latter are scentless, deep rose-coloured, with an oblong undivided lip, marked at the base of the expanded part with a deep red ring, but destitute of certain callosities remarked by Blume in his original species; at the base it is rolled up like a Cattleya, and embraces an extremely short pink downy column. For the convenience of our more scientific readers, the following transcript is added of notes made at the time of examining the plant :—

Labellum omnino læve, ungue circa columnam nanam convoluto, eique denique per spatium minimum adnato ; haud vestigium callositatis aut appendicis cujuscunque. Anthera apice biloba, valdè gibbosa, 8-locularis. Pollinia 8, per filum pulvereum colligata. Rostellum bilobum, lobis rotundis prominentibus. Glandula minuta, à rostello vix separabilis.

Messrs. Veitch inform us that this species flowers most abundantly, and that the pseudobulbs invariably have the peculiarity of producing a kind of neck about their middle; in the imported bulbs the part above the neck had all fallen off.

A third species of this genus was found on the lower ranges of the Mishmee hills by Griffith, from whom we have a dried specimen. It produces a leafy stem from two to three feet high, bearing five or six broad acuminate leaves. The flowers are few in number, at the extremity of a smooth and rather weak scape. They are somewhat larger than in *L. rosea*, with a curved spur, and an obovate four-lobed lip; their colour is unknown. Of these three species the following may be the present arrangement :—

* Column elongated.

1. *L. pauciflora* (Blume Bijdragen, 375, t. 72); floribus glabris, calcare recto, labello oblongo retuso apiculato basi bicalloso.—*Java, on Mount Salak.*

2. *L. mishmensis;* floribus glabris, calcare incurvo, labello obovato nudo obtuso apice 4-lobo.—*Mishmee* Hills Griffith.

* * Column very dwarf.

3. *L. rosea* (Lindley in Paxton's Magazine, t. 81) ; floribus villosis, calcare recto horizontali, labello oblongo obtuso nudo.—*Moulmein.*

So many species of CALANTHE, the genus nearest to Limatodes, are now in Gardens, the others are so easily procurable, and all are so very handsome, that we cannot do better than occupy a vacant space with an enumeration of such as have yet been named, distinguishing by a * those which are not yet known to be in cultivation. Three sections may be conveniently formed among them :—

* Lip spurless or nearly so.

* ? 1. C. puberula *Lindley.*—Mountains of Sylhet, where it seems to be common. Khasiya Hills (*Griffith,* no. 494).

* 2. C. gracilis *Lindley.*—Same situations as the last.

* ? 3. C. tricarinata *Lindley.*—Nepal.

* ? 4. C. brevicornu *Lindley.*—Nepal.

5. C. abbreviata *Lindley.*—Java; near the cataracts of the river Tjikundul, in the mountainous parts of Gede.

* * *Lip with a long spur ;* column much elongated.

6. C. densiflora *Lindley.*—Mountains of Sylhet.

* * * *Lip with a long spur ;* column very short.

* 7. C. clavata *Lindley.*—Mountains of Sylhet.   Khasiya Hills (*Griffith*).

8. C. angustifolia *Lindley.*—Shady mountainous places in Java, in the province of Buitenzorg (*Lobb*, 221).

9. C. curculigoides *Wallich.*—Penang and Singapore.

10. C. bicolor *Lindley.*—Japan.

* 11. C. striata *R. Brown.* (*aliàs* Limodorum striatum *Ic. Kœmpf.*, t. 2).—Japan.—Possibly this may be the same as the last, notwithstanding some apparent discrepancies.

* 12. *C. Griffithii ;* racemo laxo multifloro, ovario tomentoso, labelli lobis lateralibus linearibus obtusis intermedio subrotundo truncato denticulato sub apice dente unico magno aucto, calcare recto pendulo pubescente.—Bootan, above Telagong ; also no. 33; also " to Chuka on wet banks, 6000 feet.   Per. explanat. * * ringens."—*Griffith.*

13. C. vestita *Wallich.* (*aliàs* Cytheris Griffithii *Wight ic.* t. 1751-2.)—Burmese Empire, Mergui, Tavoy.

14. C. plantaginea *Lindley.*—Nepal and Kemaon.   Bootan, between Tussulling and Chindrippa (*Griffith*, 877).

15. C. discolor *Lindley.*—Japan ?   Java ?

16. C. parviflora ; scapo gracili multifloro pubescente, bracteis reflexis, labelli lobis lateralibus ovatis intermedio bilobo obtuso divaricato usque ad basin verrucoso, calcare glabro fusiformi pendulo sepalorum longitudine.— Java (*Lobb*, 334).

17. C. versicolor *Lindley.*—Some part of the East Indies.   Locality uncertain.

18. C. Masuca *Lindley.*—Nepal.

19. C. purpurea *Lindley.*—Ceylon.—Known from the last by its leaves being downy on the under side.

20. C. furcata *Bateman.*—Philippines.

21. C. veratrifolia *R.Brown.*—Indian Archipelago, &c.—[Var. B ; australis *Hort.*—New Holland.]

22. C. sylvatica *Lindley.*—Mascaren Islands.—[*Var. B ; natalensis *Reichenb. f. in Linnœa*, 19. 374.—Port Natal.]

Obscure species of sect. * * *.

*   C. comosa *Reichenb. f. in Linnœa*, 19. 374.—Nilgherries.

*   C. pulchra *Lindley.*—Java ; in woods on the mountains of Seribu.—Flowers pale orange.

*   C. speciosa *Lindley.*—Java ; in the deep mountain woods of the provinces of Bantam and Buitenzorg.—Flowers orange-coloured.

*   C. emarginata *Lindley.*—Java ; in the primæval woods of Mount Gede.—Flowers violet, with orange-coloured callosities on the lip.

What is the STYLOGLOSSUM of Kuhl and Hasselt, whose work on Orchids is to us completely unknown, and which is referred hither by Endlicher ?

And what can the following possibly be ?

Calanthe mexicana *G. Rehb. fil. in Linnœa*, 18. 406 ; scapo erecto foliis latis oblongis acuminatis breviore s. æquali multifloro, bracteis lanceolatis ovariis longioribus, sepalis petalisque minoribus oblongis obtusiusculis, labello ovato obtuso integerrimo puberulo calcarato, calcare tenui ovario breviore.

" This plant grows to the height of six or seven inches.   The base of the stem is covered with several leafy sheaths.   Leaves, oblong, very finely pointed, extending beyond the stem, or the same length.   The three outer calyx leaves oblong, four lines long, one broad, the two inner three lines long, one line broad, perfectly white. Lip longish, oval, blunt at the point, appearing darker coloured, covered with numerous little short hairs.   Spur very weak, pointed, somewhat shorter than the ovary.   Column short, cut quite round at the edge.   Anther at the lower end heart-shaped and notched.   Pollen masses eight, and remarkably short for a Calanthe.—Temperate Mexico.— *Leibold.*"

We repeat it, all these plants are eminently deserving of cultivation, and those which are not in England should be diligently sought for by persons living where they are found. As an encouragement to perseverance we produce the following representation of what *Calanthe vestita* was a few months since in the hands of the Messrs. Veitch.

GLEANINGS AND ORIGINAL MEMORANDA.

503. DACTYLICAPNOS THALICTRIFOLIA. *Wallich*. A climbing hardy perennial, with large yellow flowers. Native of Nepal. Introduced by Sir Charles Lemon, in 1834. (Fig. 251.)

We have before us the following memorandum concerning this plant by Mr. W. B. Booth :—" It was raised at Carclew in 1834 from some unnamed seeds which had been presented to Sir Charles Lemon, Bart. Being unacquainted with its native country, we treated it first as a greenhouse plant, but not getting it to flower, we tried it another season in the open border, planted near a standard rose-tree, the stem and branches of which afforded the necessary support to its tender and somewhat succulent climbing shoots, in which situation it flowered in September last. Root perennial, tuberous. Stem smooth, glaucous, nearly round, or but very slightly angular, of a brownish green, obscurely marked with small reddish spots. Leaves biternate—each leaf-stalk supporting for the most part nine, sometimes more, ovate acuminate leaflets—of a rich deep green above, and a pale glaucous green, with small strongly marked brownish longi-

tudinal veins beneath. All of them are furnished with a strong wiry tendril, by which the plant attaches itself to anything within its reach, and by this means attains the height of from six to eight feet. Flowers produced in clusters near the extremity of the branches, on a round slender peduncle from two to three inches long, and containing six or more pendant flowers on each. Pedicels filiform, about an inch in length. Sepals two, very small, cordate-acute, pale green. Petals four, greenish yellow, compressed. The two outer ones, alternate with the sepals, are about three-fourths of an inch long, closely connected together and conniving in such a manner as to conceal the two

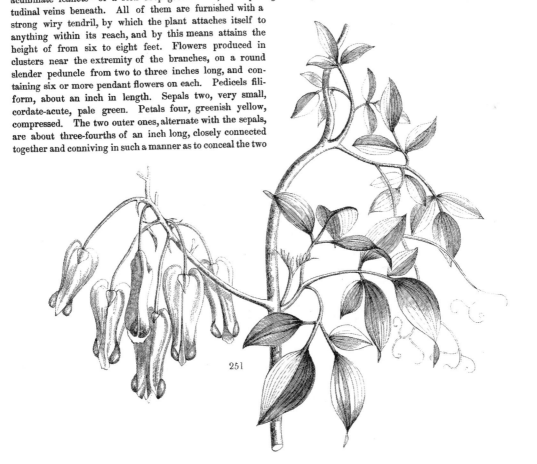

251

smaller petals and the other parts of the flower. When opened, they appear to be carinate, and exhibit the remarkable fleshy protuberance peculiar to the genus, at the base of each, and which in this species is about the length of the sepals, curved, deep green. The inner petals are curiously formed, being fiddle-shaped, broadest near the point, which is obtuse and a little elongated, and supported for half their length by a small slender thread cohering to the lower part of the stamens, and attached at the base opposite the sepals. Stamens closely surrounding the style ; the lower half of each is slightly angular and fleshy, with an uneven surface ; the upper part is capillary, bearing the anthers on the top. Style long, slender, and compressed, pale green, with a comparatively broad and thin angular point." The fruit is an oblong, cylindrical, fleshy, violet pod, about an inch long, and contains numerous kidney-shaped black seeds on two opposite parietal placentæ. Its fleshy indehiscent fruit constitutes the generic peculiarity by which it is separated from Dielytra.

504. IMPATIENS FASCICULATA. *Lamarck.* (*aliàs* Balsamina fasciculata *De Cand.; aliàs* Impatiens setacea *Colebrooke; aliàs* I. heterophylla *Wallich; aliàs* Balsamina heterophylla *Don.*) A neat succulent tender annual, with solitary axillary pale flesh-coloured flowers. Native of Ceylon.

Seeds of this pretty Balsam were sent by Mr. Thwaites, from the hilly country of Ceylon, to the Royal Gardens of Kew, where the plants blossomed in the summer of 1851. The name *fasciculata* is not a very appropriate one ; for though some of our wild specimens have the peduncles in opposite pairs, and hence appearing somewhat fasciculate, other specimens are not, and our cultivated plants had them invariably solitary in each axil. The genus or family is described as being destitute of stipules ; but in the present species, unnoticed as far as I am aware by authors, yet figured by Dr. Wight's artist, is a remarkable deflexed and very conspicuous spur at the base of each side of the leaf and decurrent with the stem, which I can look upon in no other light than as a stipule. The plant is found in a great part of the continent of India, as well as in Ceylon, appearing all over the Peninsula in marshy grounds, decorating them, as Dr. Wight says, with its large showy pink flowers. Colebrooke gathered it in Sylhet ; Mr. Griffith in Khasiya ; and Drs. Hooker and Thomson along the whole Himalayan range. Requires the same treatment, in every respect, as *Impatiens cornigera* ; and being of the same nature, will be difficult to retain as a garden plant, otherwise than by yearly importing fresh seeds from Ceylon.—*Bot. Mag.*, t. 4631.

505. PITCAIRNIA FUNKIANA. *Dietrich.* (*aliàs* Puya Funkiana *Linden.*) A charming hothouse perennial, with yellow and white spikes of flowers. Belongs to Bromeliads. Inhabits the Andes of Merida.

P. caule folioso tenuè tomentoso, foliis elongato-lanceolatis integerrimis glabris nudis nitidis, vaginis tenuè tomentosis, racemo terminali pyramidato, bracteis ovatis acuminatis calycem subæquantibus, petalis rectis apice acutis subrecurvatis basi nudis, stylo longitudine petalorum.

This beautiful plant is now in flower in the garden of M. Nauen, of Berlin. Its blossoms are white and surrounded by a calyx and bracts of a yellow colour ; it is cultivated in M. Linden's garden, in Brussels, under the name of *Puya Funkiana*, and is to be found under the same name in his catalogue (No. 5, 1850). A closer examination, however, has shown that the plant is not a Puya, but a Pitcairnia, for the former has the ovary free and not joined to the calyx, whilst the latter, as also the plant in question, has the ovary united at its base with the calyx. This species was found by Messrs. Funk and Schlim, in the deep moist valleys of the higher Andes of Merida, and was sent by them to M. Linden's establishment, and on this account we have retained the specific name proposed by the latter gentleman, in honour of the discoverer. The species belongs to the first subdivision of the genus, having its petals naked at their base, not furnished with scales. This species, like most Bromeliads, is cultivated in a hothouse, and requires a soil composed of equal parts of leaf-mould and loam, mixed with some kind of rounded sand. During the period of vegetation, plenty of water should be given, but in such a manner that all excess may run off, and therefore a layer of stones, or some such material, should be placed at the bottom of the pot. A temperature of 59° to 65·75° Fahr. is required in winter, and a more shaded or sunny place in the hothouse in summer suits this as well as other species. Bottom heat is not required, as the plant grows vigorously on the shelves of a hothouse. The plant is very handsome, and well worthy of notice. Its price is, according to M. J. Linden's catalogue, fifteen francs.—*Allgem. Gartenzeit., Oct.* 25, 1851.

506. CANNA WARCZEWICZII. *Dietrich.* A handsome hothouse perennial, belonging to the Order of Marants. Flowers scarlet. Native of Central America. Introduced by M. Von Warczewicz.

C. foliis ovatis vel ovato-oblongis cuspidato-acuminatis glabris margine cauleque coloratis, germine subgloboso papilloso colorato, calycis phyllis lanceolatis obtusis coloratis rore glauco adspersis, labio superiore corollæ limbi interioris bipartito, laciniis obverse lanceolatis obtusis, labello revoluto anguste spathulato obtuso apice emarginato, stylo lineari.

This is one of the many plants discovered by M. Von Warczewicz, who brought its seeds with him from Central America. Specimens in flower may be seen in several gardens, as, for example, in those belonging to M. Mathieu, and M. Dannenberger, of Berlin. There is no doubt that it is a new and hitherto undescribed plant. It is very beautiful, especially as the stalks, and more particularly the peduncles and pedicels, flower-bud, calyx, and bracts, are of a blood-red colour, and are covered with a bluish bloom. The flowers are bright scarlet. The plant belongs to that division of the genus which has a bifid upper lip, as in *Canna speciosa, discolor, occidentalis, compacta, carnea,* &c. —*Allgem. Gartenzeit., Sept.* 13, 1851.

507. OLEARIA PANNOSA. *Hooker.* A half-hardy evergreen shrub, native of New Holland. Belonging to the Order of Composites. Flowers white. Introduced by W. H. Fox Talbot, Esq. (Fig. 252.)

The only notice of this plant is to be found in Sir W. Hooker's *Icones plantarum*, t. 862. In that admirable collection of charming Botanical sketches, an *Olearia grandiflora*, from Adelaide in S. Australia, is figured with solitary flowerheads and large white rays. At the same time mention is made of this, as a plant found near the Murray river in South Australia; but by mistake the flowers are described as purple. In reality they are pure white, with a yellow centre. The whole plant is covered with a close white felt, except the upper side of the leaves, which are bright green and shining, with only a little cobwebby matter here and there. We presume this is not more than a greenhouse plant, among which it takes the same rank as the Canary Island Chrysanthemums.

508. BEGONIA CONCHÆFOLIA. *Dietrich.* A stove per-

252

ennial with minute red flowers, from Costa Rica. Introduced by M. Von Warczewicz. Flowered in Berlin in 1851.

B. acaulis, rhizomate repente, foliis radicalibus semipeltatis, concheato-concavis oblique ovatis angulato dentatis acuminatis basi rotundatis supra nitidis subtus albicantibus ad nervos rufo-lanatis, petiolis scapisque dichotomis coloratis rufo-lanatis, floribus dipetalis, femineis bibracteatis, capsulæ alis rotundatis, duabus angustioribus viridibus, tertia parum latiori subcrenulata colorata.

This elegant little Begonia with deep shell-like leaves has been introduced by M. Von Warczewicz from central America, and is found in many of our gardens bearing the name of B. Lindleyana, said to have been given to it by the introducer. Here is, however, some mistake, for M. Von Warczewicz himself tells us that his B. Lindleyana is one of the most beautiful of large-flowered species, whilst the present plant has very small flowers, indeed the smallest of any Begonia. It belongs to the perennial division, with a creeping many-headed rhizome, from which arise tufts of leaves and flower-stalks, but no stem ; it should consequently be placed at the end of that division. The species is characterised by its small, very elegant, shining, peltate leaves, so concave on their upper surface as to look like mussel-shells ; in this respect it differs from every other Begonia at present known, and the specific name is given to denote this peculiarity. The seeds were collected by M. Von Warczewicz, in the province of Costa Rica and in the Chiriqui-Cordilleras, during his travels in Central America, and were sent by him in 1850 to several gardeners. The plant itself is dwarfish, perennial, evergreen, and thickly covered with leaves ; it flowers in June. Its flowers are very small, but the petioles and peduncles are of a bright red colour. It requires to be kept in a hot-house, and to be cultivated in the same way as the other species. It is certainly a pretty addition to the many sorts hitherto found in cultivation. Plants can be procured from M. Bergemann of Berlin.—Allgem. Gartenzeit., Aug. 16, 1851.

509. BEGONIA STRIGILLOSA. *Dietrich.* A hothouse perennial, with rose-coloured flowers. Native of Central America. Introduced by M. Von Warczewicz.

B. acaulis, rhizomate repente, foliis oblique cordatis rubro-marginatis angulato-dentatis acuminatis, subtus ad nervos et ad marginem squamis coloratis sæpe bipartitis apice filamentosis dense obsitis supra denique subglabratis, petiolis scapisque carnosis e squamis coloratis piligeris hispidissimis, cymis dichotomis, perigonio masculo et femineo diphyllo, phyllis æqualibus, germine trialato, alis duabus obtusangulis, tertia parum latiori acutangula.

The seeds of this were sent to Europe by M. Von Warczewicz, who discovered it during his travels in Central America. According to the number in his catalogue the plant grows in the Chiriqui-Cordilleras. Like *Begonia conchæfolia*, it belongs to the perennial subdivision with creeping roots, no stems, and tufts of leaves and flower-stalks, and, notwithstanding its very different habit, must be looked upon as closely allied to them. In its hairy coating the present species has some resemblance to *B. manicata*, but the latter has a woody fleshy stem. The stalks and leaves and especially the petioles are covered with crowded, red, and often bifid scale-like hairs, much resembling the slit scales of that plant. These hairs give it a peculiar rough and wild appearance, and render it very interesting amongst many smooth and shining sorts. The leaves are obliquely heart-shaped, with a red border. The flowers, whether male or female, have only 2 sepals.—Allgem. Gartenzeit., Oct. 18, 1851.

510. CEDRONELLA CANA. *Hooker.* A handsome hardy perennial, with long interrupted spikes of purple flowers. Native of New Mexico. Belongs to Labiates.

Mr. Bentham has long ago referred the *Gardoquia mexicana* H. B. K. (*G. betonicoides* Lindl. and Graham in *Bot. Mag.,* t. 3860), to the genus *Cedronella*. The two genera are, however, in different sections of the *Labiatæ*. From that species our present one, detected by Mr. Charles Wright in an expedition from Western Texas to El Pasco, New Mexico, and number 474 of that gentleman's distributed collections, differs in the entirely glaucous stem and leaves, occasioned by a minute hoary pubescence, scarcely visible except in the recent plant, in the much smaller, more numerous, and shorter leaves, quite entire among and much below the whorls of flowers. Like that, however, the leaves abound in fragrant oil-dots. It flowers in the summer months, and makes a handsome appearance in the flower-border. Two and a half to three feet high, much branched, especially at the base ; branches opposite, square, hoary with very minute pubescence. Leaves small and entire, hoary in the upper part of the stem and near and about the flowers, and there numerous and approximate, ovate or ovato-lanceolate ; lower down larger, and cordato-ovate, or even approaching to hastate, all rather obtuse, scarcely ever acuminated, and then but slightly so, more or less strongly dentato-serrate, the teeth never reaching to the point. Whorls of flowers in axillary racemes, shortly pedunculate, the flowers pointing upwards. Calyx tubular, with five narrow, almost subulate, or subulato-lanceolate, erect teeth. Corolla almost exactly as in the *C. mexicana.*—Bot. Mag., t. 4618.

511. PEDICULARIS MOLLIS. *Wallich.* A perennial (?) herbaceous plant from the Himalaya, with long narrow whorled spikes of dull purple flowers. Belongs to Linariads. Of no horticultural interest.

Mr. Bentham well observes of this *Pedicularis,* " Species nulli proxime affinis :" the form of the corolla is extremely different from any other of the genus. It has nowhere been found except by Dr. Wallich in Gossain Than, Nepal, and in the high mountains of Sikkim-Himalaya by Dr. Hooker : from seeds sent by the latter our plants were raised in the Royal Gardens of Kew.—*Bot. Mag.*, t. 4599.

512. VANDA PEDUNCULARIS. *Lindley.* A hothouse epiphyte from Ceylon, with distichous two-lobed leaves, and brown and purple bee-like flowers. Blossoms in March. Introduced by G. Read, Esq. (Fig. 253 ; *a*, the flower slightly magnified ; *b*, the pollen-masses and caudicle.)

When enumerating the known species of Vanda at Plate 42 of our last volume, this was mentioned as not being in cultivation. On the 22nd of March of the present year, we had, however, the pleasure to receive specimens in flower from Mr. John O'Brien, gardener to G. Read, Esq.,

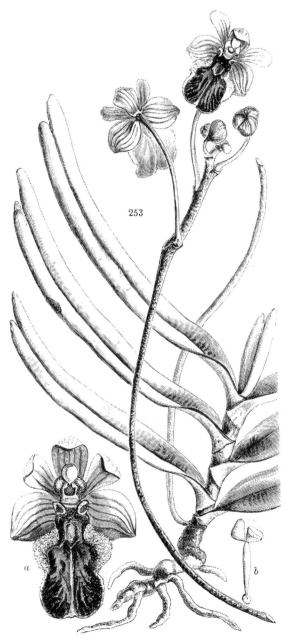

253

of Burnham, in Somerset, who had bought it as a Manilla plant at some London sale. It is in reality confined, as far as is at present known, to the island of Ceylon, where it was first found by the late Mr. James Macrae, growing on the bark of trees. The flowers are pale green, or yellowish, rather sweet-scented, with a deep purple fleshy lip bordered with green, and hairy at the edges so as to resemble some of the species of Ophrys. Growing in racemes, from six to twelve together, from the thickened ends of peduncles sometimes as much as three feet long and even furnished with side branches, these flowers wave about in the air with all the appearance of animal life, and are quite as much like hairy insects as our own wild Bee and Spider Orchises.

In some respects this is not a true Vanda ; the pollen-masses are absolutely double, and not hollowed out on one side, the caudicula is unusually long and slender, and the lip is in no degree saccate—on the contrary, it is flat, firm, and fleshy. We do not, however, at present think it expedient to separate it, whatever may happen whenever the distichous-leaved East Indian Orchids shall be thoroughly reinvestigated. In the mean while we offer the following technical description of its structure :—

" Labellum carnosum, margine tenerius, pallescens, utrinque leviter emarginatum, disco atro-purpureo, sessile, immobile, lineâ medianâ exaratâ pallidâ, in trianguli carnosi apicem desinente, basi auriculatum, etiam carnosius ; tuberculo parvo barbato inter pubem ad basin trianguli inter auriculas ; æstivatione ab apice involutum. Columna nana, erecta, tomentosa, anticè utrinque unidentata ; stigmate altè excavato circulari. Anthera 2-locularis, anticè membranacea, apiculo recurvo. Pollinia 4, geminata, aurantiaca, deltoidea, in apicem caudiculæ longæ gracilis ; glandulâ olivaceâ carnosâ subrotundâ.

513. ACROPERA FLAVIDA. *Klotzsch.* An epiphyte, with pale yellow flowers. Native of Mexico. Introduced by Mrs. Lawrence.

A. pseudobulbis ovatis, apice attenuatis, bifoliatis ; foliis oblongis, tri-quinquecostatis, acuminatis, basi longe attenuatis, utrinque nitidis ; racemis basilaribus pendulis, glabris, bracteis membranaceis, lanceolatis, acutis instructis ; perigonii foliolo supremo galeato apice apiculato recurvo ; labello vitellino ; germinibus sulcatis, scabridis pedicellisque pallide flavidis.

This plant flowered in July, 1851, in the garden

of Dr. Caspar, who received it from Mrs. Lawrence. It is probably, like *A. Loddigesii*, a native of Mexico. The pseudobulbs are two inches long ; the leaves a foot long, three inches and a half wide. The racemes consist of from eight to twelve flowers, and are from six to eight inches long. The flowers are pale yellow, with an orange-yellow lip. —*Klotzsch, in Allgem. Gartenzeit., July* 12, 1851. [The character of this species, as given by Dr. Klotzsch, is insufficient to distinguish it from *A. flavida.*]

514. LYCASTE BREVISPATHA. *Klotzsch.* An epiphyte, of the Order of Orchids, from Guatemala. Flowers pale yellowish green, with a smooth white lip. Flowered with M. Nauen, of Berlin.

L. bracteis distantibus membranaceis, aridis, fuscescentibus, superne magis inflatis, brevissime acutis, suprema ovarium vix attingente ; perianthii foliolis exterioribus oblongis, patentibus apice recurvis, intus ad basin villosulis, interioribus, basi sparsim pilosis, apice recurvis ; labello petalis breviore, trilobo, glabro, laciniis lateralibus apice truncatis, intermedia oblonga, obtusa, recurva, appendice linguæformi concava, adnata, inter lacinias laterales ; germine brevi, subincurvo, minutissime atro-punctato.

This is said to differ from *L. leucantha* in the bract being shorter than the ovary, the lip smooth, and the anterior lobes of the lip truncated. The sepals are twenty lines long, pale yellowish green ; the petals rather shorter, white, tinged with pink ; the lip shorter than the petals, whitish ; the bracts are from eight to nine lines long.—*Klotzsch, in Allgem. Gartenzeit., July* 12, 1851.

515. CERASUS ILICIFOLIA. *Nuttall.* A hardy evergreen bush or small tree. Flowers white. Belongs to Almondworts. Introduced by the Horticultural Society. (Fig. 254.)

Found in California in the first instance by Mr. Nuttall, then by Coulter, afterwards by the officers of H.M.S. Blossom, and last by Hartweg, who reports the fruit to resemble a small cherry. This is a most valuable evergreen, apparently as hardy as a Laurel, and having the foliage of a holly, with the flowers of a Bird-cherry. It has not

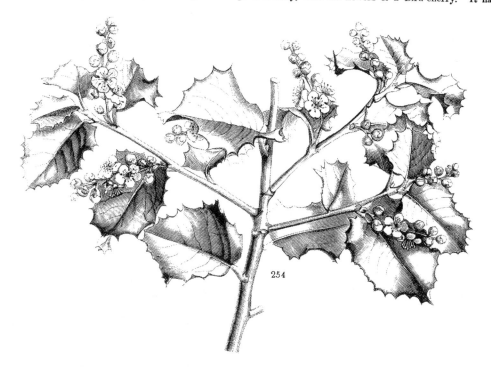

254

yet flowered, and our figure is made from a dried specimen, in order that the numerous possessors of this now not uncommon plant may know what they have to expect. It is evidently by no means so excitable as the Lauro-cerasus, and will probably stand when that species suffers.

516. NOTYLIA TENUIS. *Lindley.* (*aliàs* Notylia sagittifera *Klotzsch.*) An inconspicuous green-flowered Orchidaceous epiphyte. Native of Demerara. (Fig. 255.)

We copy the annexed figure from the *Icones plantarum rariorum Horti Botanici Berolinensis* of Link, Klotzsch, and Otto ; it represents very well the lower part of the long slender spike of a plant of mere botanical interest, which Dr. Klotzsch believes to be the *Pleurothallis sagittifera* of Humboldt.   He says that he has compared it with the original specimens, and is unable to perceive the least difference. To this we can offer no objection ; but we are obliged to add that it is most certainly not the *N. punctata* of the genera and species of Orchidaceous plants, as the learned Prussian supposes ; differing in its long slender spike, and in the absence of that callus on the lip by which *N. punctata* is so strikingly characterised. As the species seem to be scarcely at all understood by foreign botanists, we may as well take the present opportunity of enumerating those we are acquainted with. They are much alike, but vary in the form of their lip, in the size of their flowers, in the presence or absence of a lip-tubercle, and in similar points.

255

1. N. punctata *Lindl., in Bot. Reg.*, 930, 1825 ; (*aliàs* Gomeza tenuiflora *Loddiges; aliàs* Pleurothallis punctata *Ker.*)

2. N. tenuis *Lindl., in Bot. Reg.*, 1838; (*aliàs* Pleurothallis sagittifera *H.B.K.*, according to Klotzsch, and therefore N. multiflora *Lindl.* ; and probably the plant figured under that name by Sir W. Hooker in the *London Journal of Botany*, vol. iii., t. 10.)

3. N. incurva *Lindl., in Bot. Reg.*, 1838, misc. 167.

4. N. Barkeri *Id.*, 1838, misc. 168.

5. N. micrantha *Id.*, 1838, misc. 170.

6. N. pubescens *Id.*, 1842, misc. 72.

7. N. aromatica *Id.*, 1841, misc. 77.

8. N. bicolor *Id., in Plant. Hartweg.*, no. 93.*

9. N. trisepala *sp. nov.* ; racemo gracili tenui ascendente, bracteis ovario brevioribus, sepalis clausis disjunctis, labello subhastato ecalloso.—Formerly received from M. Van Houtte ; native country unknown. The flowers are almost white.

10. N. Tridachne (*aliàs* Tridachne virens *Liebmann*) ; sepalis lateralibus omnino connatis labello trullæformi acuminato basi angustato ecalloso.—This was received by the Horticultural Society, from Mr. Weilbach of Copenhagen ; and is remarkable for the combination of its sepals into two, instead of three ; the petals are yellow, with one or two pale orange bands ; the lip is clear yellow.

It is probable that *N. Hügelii*, of *Fenul Nov. gen. et sp. plant.* p. 3, a work received since this was printed, is the same as the present species. We find it thus defined from a Mexican specimen in the garden of Baron Charles v. Hugel :—" Pseudobulbi lineari-oblongi compressi minuti. Folia solitaria coriacea linguæformia planiuscula, subtus basi carinata. Racemus radicalis pendulus multiflorus, pedicellis bractea subulato-setacea triplo, flore triente longioribus recurvo-patulis. Perigonii viridi-flavi foliola externa lateralia labello supposita in unum *apice integerrimum* coalita, cum superiore sublongiore lineari-lanceolata acuta ecarinata navicularia apice recurva ; interna subbreviora ac dimidio angustiora lineari-subfalcata acuminata medio superposite aurantico-bi-v. quadripunctata. Labellum porrectum unguiculatum trullæforme integerrimum acutum apice subincurvum."

In addition to these, others no doubt remain unexamined among South American collections, exclusive of the following :—

Doubtful Species.

11. N. laxiflora *Westcott, in the Phytologist,* i. 54 ; perhaps the same as *N. aromatica.*

12. N. orbicularis *Richard & Galeotti ;* a Mexican plant that we have never seen.

517. KLUGIA NOTONIANA. *De Candolle. (aliàs* Wulfenia Notoniana *Wallich ; aliàs* Glossanthus Notoniana *Brown ; aliàs* Glossanthus malabarica *Klein ; aliàs* Glossanthus zeylanica *Brown.)* An annual weedy plant, with deep blue flowers. Native of Ceylon. Belongs to Gesnerads.

The genus Klugia of Schlechtendahl in *Linnæa* (1833), the same with Glossanthus of Klein (1835) and of Brown, was founded on a Mexican plant ; but a congener, if not congeners, are found in India : the present is one of them, remarkable for the great obliquity of the base of the leaf, and the brilliant colour of the blue flowers. Our living plants were received from Ceylon, through the kindness of our valued friend Mr. Thwaites, of the Botanic Gardens, Peradenia. Hence we suspect it may be the *Glossanthus zeylanica* of Mr. Brown, l.c., without description. It is, however, certainly the *Wulfenia notoniana* of Dr. Wallich, and consequently *Glossanthus notoniana* of Mr. Brown, and *Klugia notoniana* of De Candolle, whose name we here adopt. It is abundant in the Neilgherry hills, and flowers in the stove in September. A soft-stemmed tropical plant, of low decumbent habit, and producing roots from the under side of the stem. It is at this time growing and flowering freely in a warm stove. A mixture of light loam and peat-soil suits it, and it appears to love moisture ; it is, however, liable to suffer by an excess of moisture in the atmosphere of the house in the winter, and more particularly towards the spring, as by that time its powers have become exhausted and it is apt to damp off.—*Bot. Mag.,* t. 4620.

518. ACANTHOSTACHYS STROBILACEA. *Klotzsch. (aliàs* Hohenbergia strobilacea *Schultes.)* A curious perennial, with very narrow spiny leaves, like the Pine Apple, and a short prickly cone of yellow flowers in orange-coloured bracts. Belongs to Bromeliads. Native of Mexico. (Fig. 256.)

According to Mr. Otto this comes from the southern provinces of Brazil, where it was first found by Martius, and afterwards by Sello. It flowers in the stove in June and July, in equal parts of sand and decayed vegetable mould. A second species is *Hohenbergia (Acanthostachys) capitata,* also from Brazil. One of the great peculiarities of this genus is its having its ovules in pairs only, and not in crowds on the edges of an axile placenta ; it is inferior-fruited, like Ananassa itself. The leaves are very long and narrow, thick, curved, prickly, channelled, and scurfy. The scape is long, simple, mealy, and bears at the base of the prickly spike (or cone) a pair of very long channelled leafy spathes.— *See Link, Klotzsch, and Otto's Icones.*

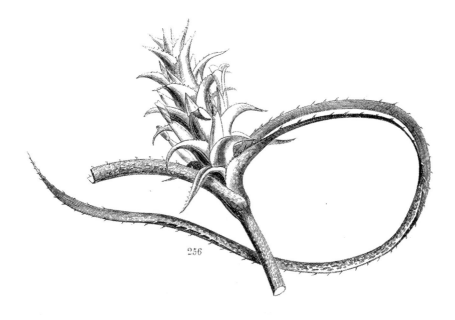

256

PLATE 82.

LConstans del.& zinc.

Printed by C.F.Cheffins, London.

[PLATE 82.]

THE DARK PURPLE HELLEBORE.

(HELLEBORUS ATRORUBENS.)

———◆———

A hardy Herbaceous Plant, from CROATIA, *belonging to the Order of* CROWFOOTS.

═══════════════

Specific Character.

THE DARK PURPLE HELLEBORE. Radical leaves quite smooth, pedate, pale beneath and shining : those of the stem nearly sessile and palmate. Stem rather angular, branched by bifurcation. Sepals roundish, coloured.

HELLEBORUS *ATRORUBENS;* foliis radicalibus glaberrimis pedatisectis, subtus pallidioribus nitidis, caulinis subsessilibus palmatipartitis, caule subangulato bifide ramoso, sepalis subrotundis coloratis.—*D. C.*

Helleborus atrorubens : *Waldstein & Kitaibel, Plantæ rariores Hungariæ*, vol. iii., p. 301, t. 271; *De Cand. Prodrom.* i. 47.

═══════════════

Although very far from a novelty, this curious plant is hardly known in the gardens of this country. That which is now figured was kept in a greenhouse in the Garden of the Horticultural Society, where it had been received from Mr. Van Houtte. It is, however, perfectly hardy, flowering in March and April in a border among shrubs. The skirts of a clump of Rhododendrons suit it perfectly.

It was first made known to botanists by Waldstein and Kitaibel, who give a very bad figure of it in their great work on the plants of Hungary, in which they state that it is found wild in woods and thickets in Croatia, in great abundance near Korenicza.

Although a native of such a country, in which the winter's cold and the summer's heat are far beyond anything experienced in these islands, the plant is much more beautiful in a greenhouse than in the open air. It is only in the former, indeed, that its peculiar and rather striking tints become developed. The leaves are liable to considerable difference of form, being five-parted, or even nine-parted, but they never assume the lobed condition of the other purple species *H. purpurascens*, nor are the lobes united half-way up; on the contrary, with the exception of the side divisions, they are distinct almost to the very base. The stem is about eighteen inches high, and produces its branches by two or three series of forkings. The flower-buds are a deep black-purple; the expanded flowers are of a peculiar violet-purple, except at the edges and centre, both which are green; but in a few days the violet flies off, and leaves nothing behind except a dingy green tinted with dull purple. No such brilliancy as is found in our figure is produced in the open air as far as we have remarked. The plant is, however, perfectly hardy.

The Honourable W. F. Strangways, who has paid much attention to the species of this genus, has favoured us with the following useful memorandum respecting them :—

Since I find that HELLEBORES are attracting some notice as fine hardy herbaceous plants, fit for undergrowth in woods and shrubberies, the following synopsis may perhaps be acceptable :—

A. *Suffrutescent, with biennial stems.*

H. argutifolius
    lividus   } three-leaved.

H. fœtidus   palmate-leaved.

B. *Herbaceous, with annual stems.*

H. niger, two or three varieties
    abchasicus
    olympicus   } with coloured flowers.
    orientalis
    atrorubens

H. cupreus
    purpurascens   } with dusky flowers.
    intermedius

H. viridis
    laxus
    pallidus
    odorus   } with green flowers.
    angustifolius
    graveolens

H. Bocconi, and perhaps another species—doubtful—in Italy. H. fœtidus is a native of Wales ; H. viridis, of Dorsetshire ; H. argutifolius and lividus, of Corsica ; H. niger of the Alps ; H. abchasicus, orientalis, and olympicus, of the Levant. The rest, of Hungary. All, except lividus, of the easiest culture in shady situations.

PLATE 83.

L.Constans del.& zinc.

Printed by C.F.Cheffins, London.

[PLATE 83.]

THE CILIATED RHODODENDRON.

(RHODODENDRON CILIATUM.)

———◆———

A hardy (?) Evergreen Shrub, from SIKKIM-HIMALAYA, *belonging to the Order of* HEATHWORTS.

Specific Character.

THE CILIATED RHODODENDRON. A low rigid shrub. Branches, leaf, and flower-stalks covered with stiff spreading hairs. Leaves on short footstalks, elliptical, obovate, very sharp, bright green above, the margins and mid-rib with stiff spreading hairs, paler and rather glaucous below, dotted with small scales. Flowers four or five together, pale purple, on stout short flower-stalks. Sepals broadly ovate, blunt, ciliated on the margin. Corolla bell-shaped, with spreading recurved lobes. Stamens ten. Ovary scaly, five-celled.—*J. D. Hooker.*

RHODODENDRON *CILIATUM;* humile, suffruticosum ; ramis petiolis pedicellisque rigidè villosis, foliis subsessilibus ellipticis obovatis acutissimis laetè viridibus ciliatis subtus pallidis glaucescentibus minutè lepidotis, floribus 4—5-nis pallidè purpureis, pedicellis brevibus rigidis, sepalis latè ovatis obtusis ciliatis, corollâ campanulatâ patentissimâ imo margine recurvâ, staminibus 10, ovario lepidoto 5-loculari.

———

Rhododendron ciliatum : *J. D. Hooker, Sikkim Rhododendrons,* t. 24 ; *Journal of Horticultural Society,* vol. vii., pp. 77, 95 ; *Botanical Magazine,* t. 4648.

———

THIS is the first of the true Sikkim Rhododendrons which has flowered in this country. Messrs. Standish & Noble exhibited the specimen now represented to the Horticultural Society in the beginning of last March, and the species has also produced its flowers at Kew. It is not a little remarkable that neither of them resembled in colour the beautiful figure in the Sikkim Rhododendrons, or indeed each other. In a wild state the blossoms appear to be violet; with Messrs. Standish and Noble they were pale delicate rose-colour; at Kew they were almost white.

This is, no doubt, one of the most cultivable of the Indian alpine species, those who have had the worst success with others having managed to keep it in health. It has a peculiarly bright green aspect, breaks its buds very early if in a greenhouse, and seems as little impatient of confinement as of external cold when exposed. It does not appear to grow above a foot or two high, and begins to blossom when not more than six inches tall. The flowers themselves are delicate and beautiful, but the great value of the plant may be expected to consist in its giving dwarfness to mules with the tall and hardy Rhododendrons, such as *ponticum, catawbiense*, and *maximum*. Dr. Hooker, in his very able and instructive paper on the climate of the Sikkim Himalaya, in the Journal of the Horticultural Society, speaks thus of the plant before us :—

" R. *ciliatum*.—Distribution and range : *Sikkim*—9000 to 10,000 feet—in rocky valleys of the interior.

" This forms a small very rigid shrub, growing in clumps 2 feet high, generally in moist rocky places. Odour faintly resinous and pleasant. Corolla $1\frac{1}{2}$ inch long, nearly as much across at the mouth ; tube rather contracted below, limb 5-lobed, colour pale reddish-purple ; upper lobe obscurely spotted. Allied to R. *barbatum*, but widely different in stature, habit, and the scattered scales on the under surface of the leaves. I have not observed it in other valleys than those flanked by snowy mountains, where it is common, scenting the air in warm weather. The scales (as in its congeners) are orbicular, sessile, attached at the centre, formed of 3 concentric series of cells surrounding a central one, in which a resinous fragrant oil is secreted."

PLATE 84.

L.Constans del.& zinc.

Printed by C.F.Cheffins, London.

[PLATE 84.]

THE DARK-EYED FRINGED DENDROBE.

(DENDROBIUM FIMBRIATUM; VAR. OCULATUM.)

———◆———

A Stove Epiphyte, of great beauty, from the EAST INDIES, *belonging to* ORCHIDS.

===============

Specific Character.

THE FRINGED DENDROBE. Stems terete, leafy. Leaves ovate-lanceolate. Racemes lateral, lax, many-flowered. Bracts herbaceous, minute. Sepals oblong, spreading flat. Petals larger, toothletted. Lip undivided, rounded, hooded, shaggy, fringed; the fringes lacerated.

Var. B. *Dark-eyed.* Flowers larger, with a deep brown spot in the middle of the lip.

DENDROBIUM *FIMBRIATUM* (STACHYOBIUM); caulibus teretibus foliosis, foliis ovato-lanceolatis, racemis lateralibus laxis multifloris, bracteis herbaceis minutis, sepalis oblongis patentissimis, petalis majoribus denticulatis, labello indiviso rotundato cucullato villoso fimbriato; fimbriis laceris.

Var. B; *oculatum*, floribus majoribus, labelli medio piceo aterrimo.

———

Dendrobium fimbriatum : *Hooker, Exotic Flora,* t. 71 ; *Lindley, Genera & Species,* no. 38. Var. B ; D. fimbriatum oculatum : *Botanical Magazine,* t. 4160.

———

OF this most beautiful plant our gardens contain two distinct varieties; one with whole-coloured flowers; the other with a deep rich pitch-brown spot in the middle of the lip. In both, the colour is otherwise of a rich apricot-yellow, rendered the more brilliant in consequence of the surface and edge of the lip being cut up into glittering points innumerable. The first was sent home many years ago by Dr. Wallich, and flowered in the Botanic Garden at Liverpool about the year 1822; the second, now figured from Chatsworth, is of much more recent introduction, and is sometimes known under the erroneous name of *Paxtoni*, which is a two-flowered species. The wild specimens in our possession, belonging to the second or dark-eyed form, were collected by Griffith in Mergui; the whole-coloured form seems to come only from Nepal.

It is most nearly related to *D. clavatum* (our Fig. 189), which is readily known by its long membranous bracts, and from *D. Gibsoni* (our Fig. 204), the flowers of which are smaller, and never open flat; neither of those species has any fringes upon the petals.

A CATALOGUE

Of the Dendrobes *belonging to the Section* Stachyobium, *having an undivided lip; with their synonymes and horticultural merits.*

Group 1.—AUREA.

1. D. auriferum *Lindley.*—China.—Flowers yellow, with long tapering points, and enclosed in the hooded bracts of short lateral racemes. Only known from a drawing in the library of the Horticultural Society.

2. D. flavescens *Lindley* (aliàs *Onychium flavescens* Blume).—Java.—Flowers small, yellow.

3. D. rhombeum *Lindley.*—Manilla.—Very like *D. aureum*, but the flowers are racemose.

4. D. sulcatum *Lindley.*—East Indies.—Flowers erect, yellow, whole-coloured.

5. D. polyanthum *Wallich.*—Moulmein.—Flowers yellow (?) very pale (?).

6. D. Gibsoni *Paxton.*—East Indies.—Flowers in long pendulous racemes, rich apricot-yellow, with a purple stain on the lip.

7. D. fimbriatum *Hooker.*—Nepal, Burma.—Var. A; flowers rich orange-yellow, whole-coloured: var. B; flowers larger, with a rich purple-brown centre to the lip.

8. D. clavatum *Wallich.*—Assam.—Flowers large, bright yellow, with a double rich brown stain in the middle of the lip.

9. D. moschatum *Wallich* (aliàs *D. Calceolus* Hooker; aliàs *D. cupreum* Herbert).—Burma, Ava, Pegu.—Flowers large, pale nankeen-coloured, richly stained and veined with crimson, musky-scented.

Group 2.—AXANTHA.

10. D. Dalhousieanum *Paxton.*—East Indies.—A magnificent plant, with large cream-coloured flowers tinged with rose, and a pair of broad purple blotches on the lip.

11. D. formosum *Roxburgh.*—East Indies.—Flowers white, very large.

12. D. mutabile *Lindley* (aliàs *Onychium mutabile* Blume).—Java.—Flowers pale rose; lip with three yellow glands.

13. D. sclerophyllum *Lindley* (aliàs *Onychium rigidum* Blume).—Java.—Flowers whitish, with three yellow glands on the lip.

14. D. triadenium *Lindley.*—Java.—Flowers nearly white; with a violet spot on the ends of

the sepals and lip: the latter with three yellow glands. Possibly these three last may be only varieties of each other.

15. D. aduncum *Lindley*.—East Indies.—Flowers almost transparent, of the most delicate pink.

16. D. japonicum *Lindley* (aliàs *Onychium japonicum* Blume).—Japan; cultivated in Java.— Flowers lilac, sweet-scented, with a ciliated lip.

17. D. nudum *Lindley* (aliàs *Onychium nudum* Blume).—Java.—Flowers pale purple, changing to yellow.

18. D. calcaratum *A. Richard*.—Island of Vanikoso.

19. D. ramosum *Lindley*.—East Indies.—Flowers small, colourless.

20. D. herbaceum *Lindley*.—East Indies.—Flowers small, greenish, inconspicuous.

N.B.—D. cassythoides *A. Cunningham*, a leafless creeping plant from Port Jackson, described in the Botanical Register for 1836 under figure 1828, the pollen of which is unknown, is probably some Vanilloid plant allied to Cyrtosia, if not belonging to that genus.

GLEANINGS AND ORIGINAL MEMORANDA.

519. ILEX PERADO. *Hort. Kew.* (*aliàs* I. platyphylla *Webb & Berthellot.*) A hardy evergreen tree, with broad flat foliage, and bright red fruit. Native of the Canary Islands. Flowers white in June. (Fig. 257.)

An old inhabitant of our greenhouses, but to all appearance perfectly hardy near London. The first published account of it is to be found in *Plukenet's Almagestum* (t. 262), where it is represented under the name of " Aquifolium amplissimis foliis ex insulis Fortunatis." In the first edition of the *Hortus Kewensis* it was placed among other Hollies as *Ilex Perado*, by which designation it was universally known, until Messrs. Webb and Berthellot called it *I. platyphylla*, supposing *Ilex Perado* to be the same as the *I. maderensis* of Lamarck, for which we find no sufficient authority. The Perado of Kew was a garden plant, and has descended to our days in the form which is now represented. According to the learned authors of the Natural History of the Canaries, this plant grows in the dense forest of Agua Garcia in the Canaries, where it forms a pyramidal tree twenty feet high, and is called Naranjero Salvage. They believe it to be strictly a Canary plant, and not to be known in Madeira. In gardens the species resembles a broad, flat, roundish-leaved Holly, with little or no toothing on the margin. The flowers are white, numerous, much larger than in *I. aquifolium*, and are succeeded by bright red spherical berries. It is a truly noble evergreen.

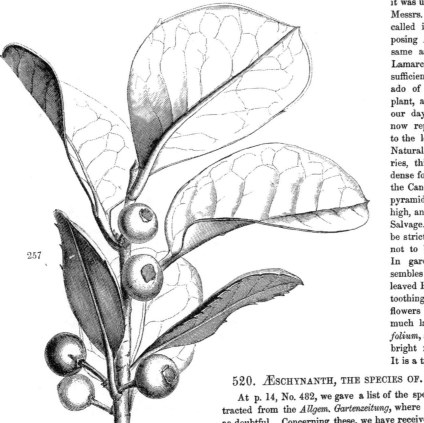

257

520. ÆSCHYNANTH, THE SPECIES OF.

At p. 14, No. 482, we gave a list of the species of this genus extracted from the *Allgem. Gartenzeitung*, where two kinds were named as doubtful. Concerning these, we have received the following memorandum from Mr. Moore, of the Apothecaries' Garden, Chelsea.

521. ÆSCHYNANTHUS DISCOLOR. Leaves elliptic, acuminate, obsoletely sinuate-dentate, glabrous, fleshy, veinless.

Flowers axillary, with pentagonal solitary or twin peduncles, calyx glabrous; the tube one-third as long as the subulate segments, which are one-third shorter than the glabrous corolla; limb of the corolla ciliate; stamens much exserted, hairy; style very short, included within the tube of the calyx.

A glabrous shrub with round greenish stems and thick broadly lanceolate acuminate stalked leaves, without evident veins, but having beneath a prominent purple costa, and a corresponding channel above; the margin almost entire when fully grown, but having a few glandular teeth-like projections when young; four inches long, an inch and a half broad, on petioles half an inch long, deep green above (sometimes obscurely dotted with dull purple), dull reddish purple beneath. Flowers axillary, with minute bracts at the base of the glabrous distinctly pentagonal peduncles, which are longer than the petioles. Tube of the calyx prismatical, nearly as long as the peduncles, and three times shorter than the subulate segments of its limb, glabrous and purplish throughout. Corolla one-third longer than the calyx, the green tube widening upwards slightly curved, the limb oblique with roundish ciliated segments, marked within with three converging chocolate-brown bars, which meet within the border and form an angular figure on each segment. Stamens half as long again as the corolla, hairy above. Style half as long as the tube of the calyx, straight, the stigma forming a groove at the scarcely expanded apex. Ornamental owing to its coloured foliage. This is the _Æ. atrosanguineus_ Hort. (not of Paxton's Bot. Dict., which is stated to have dark red flowers). It may also be the _Æ. atropurpureus_ Hort. Van Houtte (Walp., Rep. V., 521), but the leaves in our plants are scarcely spotted except by accidental discoloration, and the flowers of Van Houtte's plant are not described.

522. ÆSCHYNANTHUS MARMORATUS. Leaves oblong-lanceolate (or obovate-lanceolate or ovate), acuminate, scarcely toothed, obscurely veined; flowers axillary, calyx puberulous, the tube obsolete; segments of the limb subulate-aristate; corolla glabrous, twice as long as the calyx; the limb ciliated; stamens exserted, hairy above; style nearly equalling the tube of the corolla, densely villous.

A smooth shrub with round green stems and broadly lance-shaped acuminate variable leaves which are fleshy, the obscure veins pallid green on both sides, with deep green intervening above, and reddish purple below; they are three and a half inches long, and an inch and a half broad, stalked, and obsoletely glandular-toothed when young. Flowers axillary, on pentagonal peduncles as long as the petioles. Calyx clothed with scattered hairs, divided almost quite to the base, the segments subulate-aristate, purplish, and about half as long as the corolla. The corolla has a curved tube widening upwards, and an oblique limb of roundish ciliated segments, the tube green, the limb blotched with chocolate-brown. Stamens much exserted, hairy in the upper part. Style nearly as long as the tube of the corolla, thickened and glabrous below, densely villous above, terminated by an expanded transversely grooved stigma. The marbled leaves give the plant an ornamental character.

This is the _Æ. zebrinus_ of English gardens, and is probably the _Æ. zebrinus_ Hort. Van Houtte (Walp., l.c.) It cannot, however, be the _Æ. zebrinus_ of Paxton's Bot. Dict., for that is stated to have scarlet flowers.

Both this and _Æ. discolor_ are evidently nearly related to _Æ. purpurascens_ Hasskarll; but, independently of other differences in the foliage and flowers, _Æ. discolor_ is at once distinguished by its very short style, and _Æ. marmoratus_ by its obsolete calyx-tube.

523. BESCHORNERIA TUBIFLORA. _Kunth._ (_aliàs_ Fourcroya tubiflora _Kunth & Bouché._) An Aloe-like greenhouse perennial, belonging to Amaryllids. Flowers greenish-brown. Native of Mexico.

Imported from Mexico to the Royal Gardens of Kew, where it produced its Agave-like blossoms in a cool greenhouse in February 1852. Professor Kunth considers the genus to be intermediate between Littæa (Agave, sect. 2) and Furcræa, differing from the latter in habit, from the former in its included stamens, and from both in the tubular flower. Stemless. Leaves radical, tufted, spreading and more or less recurved, linear, sword-shaped, very much acuminated, eighteen inches to two feet long, thickened and narrowed and triangular at the base, minutely striated, glaucous-green, beneath rough to the touch, and when seen under the microscope muricated on the nerves, and sharply denticulated at the margin. Scape erect, in our plant four feet high, bearing a many-flowered erect raceme. Flowers fascicled, drooping two to four from the top of a blunt tooth or swelling, bearing a large purple-coloured, ovate, membranaceous bractea. Pedicels shorter than the bractea, green, terete, bearing each a subulate bracteole at its base. Perianth divided to the top of the ovary, into six green, spathulate, nearly equal segments (brownish-purple externally), erect and approximating into a tube, the apices only spreading. Stamens six, equal, erect, rather shorter than the perianth; filaments subulate; anthers linear-oblong, pale green. Style dilated and six-angled at the base; stigma small, three-lobed.—_Bot. Mag._, t. 4642.

524. ECHINOCACTUS LONGIHAMATUS. _Galeotti._ A glaucous ribbed succulent plant with very long recurved spines and large yellow flowers. Native of Mexico. Blossoms in July.

A fine and handsome species :—remarkable in the very prominent ridges, the large and regularly arranged spines, the central one very long, flattened, and hooked at the end, and handsome in the size and colouring of its flowers, both in the bud and when fully expanded. It is a native of Mexico, and appears to have been introduced to our collections by M. Galeotti.—_Bot. Mag._, t. 4632.

525. BERBERIS TRIFURCA. A hardy (?) evergreen shrub, with pinnated leaves. Flowers unknown. Native of China. Introduced by Mr. Fortune. (Fig. 258.)

B. (Mahonia) *trifurca;* foliis pinnatis, foliolis ovato-elongatis juxta basin distanter spinoso-dentatis apice sæpissime altè tridentatis terminali sessili angustiore et longiore.

Mention of this curious species of pinnated Berberry is made by Mr. Fortune in his recent account of a visit to the Tea Countries of China, and is alluded to under the name of *B. trifurca*, among the remarks upon *B. nepalensis* at Plate 79. Although undoubtedly allied to the latter species, it is readily distinguished by its long leaflets, which have a few coarse toothings near the base, then a long toothless interval, and at the point three stout teeth ; in addition to which the *terminal leaflet is sessile.* Living plants exist in the nursery of Messrs. Standish and Noble, of Bagshot, to whom it was sent by Mr. Fortune.

526. HELMIA RACEMOSA. *Klotzsch.* A climbing shrubby hothouse plant from Central America. Belongs to the Order of Yams. Flowers small, yellow and purple. Introduced by Von Warczewicz.

H. suffrutex volubilis, glaber ; rhizomate tuberoso, carnoso ; ramis

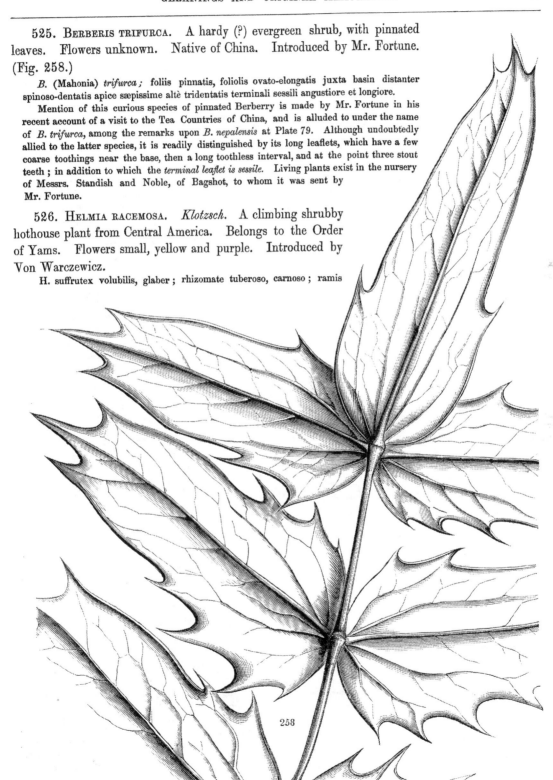

258

teretibus ; foliis sparsis, cordato-ovatis, acuminatissimis, 9-nerviis, supra læte viridibus, subtus pallidis, nitidis, versus basin glandulis scrobiculatis sparsis obsitis ; petiolis basi apiceque incrassatis ; racemis masculis axillaribus solitariis, racemosis, folio subtriplo brevioribus ; rachibus angulato-alatis, alis brevissimis, minutissime puberulis ; floribus solitariis bracteisque duabus ad basin pedicelli purpureis ; perigonii turbinato-rotati laciniis ovato-oblongis, subobtusis, patentissimis ; staminibus 3 brevissimis, arcte approximatis, centro disci atro-purpurei hexagoni insertis ; antheris introrsis bilocularibus post dehiscentiam saturate aureis ; rudimento stylino nullo.

Although it is not easy to class a diœcious plant, of which only one sex is known, and that with scarcely characters enough to authorize its being made the basis of a new genus, yet I think that in the present case there are two characters sufficient for this purpose ; viz. the presence and form of the sheath of the male flowers, and the presence of two unequal great bracts at the base of each peduncle. As to the species, there is the cylindrical twining stem, the thickness of a crow-quill. The petiole, also cylindrical, two to three inches long, and thickened both at the top and the bottom. The leaf oval and heart-shaped, with a long pointed apex, nine-nerved, three to three and a half inches broad, and three to four inches long, counting from the insertion of the petiole to the point, whilst the two side lobes at the base project six to nine lines beyond the point of insertion. Besides, hitherto no species of Helmia has been found which has an axillary raceme as short compared with the size of the leaves as that of the plant now in question. Upon this lie the dark yellow-red and carmine corollas. The corolla is as long as the calyx, as are also the style and stamens, which latter have red filaments and bluish anthers. The bulbs of this plant are similar in shape and size to that of *Dioscorea alata,* and were received by M. Mathieu of Berlin, who has them now (December 1851) in flower, from M. Von Warczewicz, who discovered them in Central America.—*Klotzsch, in Allgem. Gartenzeit., Dec.* 13, 1851.

527. CESTRUM BRACTEATUM. *Link & Otto.* (*aliàs* C. stipulatum *Vellozo.*) A greenflowered, greenhouse shrub. Native of Brazil. Belongs to Nightshades. (Fig. 259.)

This species is remarkable for the large size of its greenish bracts, which extend from the calyx as far as the limb of the corolla. It forms a stout branching shrub, five to six feet high, with green downy branches. The leaves are pale green, lanceolate, wavy, with rather conspicuous veins, and bear at their base a pair of roundish green ears, which have been called stipules by Graham, and the scales of axillary buds by Dunal. The flowers are slightly downy, pale green, in short spikes or fascicles, and when young are concealed by the great downy glumaceous bracts in which they are enveloped. According to Dunal the species inhabits the open deciduous forests of Brazil. Being as destitute of odour as of colour it is of little horticultural interest.

259

528. BEGONIA PUNCTATA. *Link, Klotzsch, & Otto.* A hothouse perennial with panicles of pink flowers. Belongs to Begoniads. Native of Mexico. (Fig. 260.)

A handsome stemless herbaceous plant, with a creeping rhizome. The leaves are cordate, cut into about seven toothed palmate lobes bordered with fine bristles, slightly hairy on each side, dark green on the upper, pale green on the under side, with a tinge of red towards the edge ; their stalks are furrowed, covered with spreading hairs, and furnished with a purple ramentaceous collar just beneath the lamina. The sepals are in pairs, oblong, a little narrowed to the base, bright rose-colour, with deep red spots on the outside. Fruit dotted with scarlet, the wings rounded, one being very large and bright rose-colour. This plant, figured in *Link, Klotzsch, & Otto's Abbildungen,* is very near B. *heracleifolia* and *crassicaulis.* The former differs in having clear green, smaller, and more deeply cut leaves, crenate bracts, and unspotted flowers ; the latter in blossoms without leaves, white flowers, small oblong leaves, and perfectly circular sepals. They all have the same great double placenta, and belong to the section (?) Diploclinium.

529. RYTIDOPHYLLUM HUMBOLDTII. *Klotzsch.* (*aliàs* Gesnera Humboldtii *Warcz.*) A hairy half-shrubby hothouse plant, with greenish flowers spotted with purple. Belongs to Gesnerads. Native of Central America. Introduced by M. Warczewicz.

Rytidophyllum Humboldtii; suffruticosum; caule erecto, ramoso, villoso; foliis oblique oblongis, membranaceis, inter se inæqualibus, grosse serratis, acuminatis, basi subattenuatis, supra dense pubescentibus, subtus petiolisque villosis; corymbis in apice ramulorum axillaribus, longe pedunculatis, 2—3-floris, foliosis; calycis laciniis ovato-lanceolatis, acuminatis, 5-nerviis, utrinque germineque villosis, tubo corollæ hirsutissimo subæquantibus; corollæ lobis patentibus, rotundatis, extus evanescente pubescentibus, intus glabris, virescentibus, purpureo-maculatis; filamentis superne sparsim glanduloso-pilosis; stylo scabro; stigmate incrassato; disco epigyno annulari, 5-crenato.

A half shrubby plant, about three feet high, collected by M. von Warczewicz in Veragua

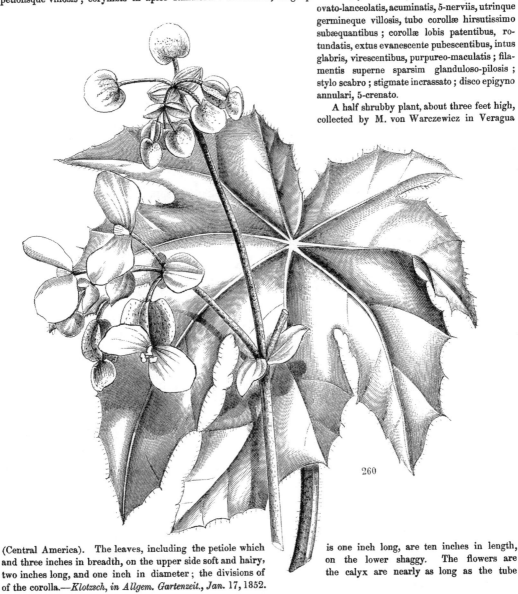

260

(Central America). The leaves, including the petiole which is one inch long, are ten inches in length, and three inches in breadth, on the upper side soft and hairy, on the lower shaggy. The flowers are two inches long, and one inch in diameter; the divisions of the calyx are nearly as long as the tube of the corolla.—*Klotzsch, in Allgem. Gartenzeit., Jan.* 17, 1852.

530. RYTIDOPHYLLUM TIGRIDIA. *Klotzsch.* (*aliàs* Gloxinia Tigridia *Ohlendorff; aliàs* Sisyrocarpum Ohlendorffii *Klotzsch.*) A climbing half-shrubby hothouse plant, from Venezuela. Flowers very large, greenish, spotted with purple. Belongs to Gesnerads. Introduced by M. Moritz.

Rytidophyllum Tigridia; suffruticosum; caule scandente, teretiusculo, hirsutissimo; foliis oblique ellipticis, inter se inæqualibus, petiolatis, grosse serratis, ciliatis, supra sparsim pubescentibus, subtus nervoso-hirsutissimis ; corymbis in apice ramulorum axillaribus, longissime pedunculatis, folio subtriplo longioribus, 4—5-floris ; calycis laciniis ovatis, acutis, obsolete trinerviis, utrinque germineque dense pilosis, tubo corollæ lanato, 4 duplo brevioribus, post anthesin patentibus ; corollis maximis, virescentibus, purpureo-maculatis, limbi lobis patentibus, rotundatis, intus glabris ; filamentis styloque hirtis ; stigmate incrassato ; seminibus longissimis, scobiformibus ; disco epigyno annulari, obsolete 5-crenato.

This climbing bush, of which plants were sold for 6s. each by Messrs. Ohlendorff & Sons, in Hamburgh, in 1845, was discovered by M. C. Moritz in the snowy mountains of Merida (Venezuela), who sent seeds of it to the above-named gentlemen ; dried specimens of it, collected by Mons. Bonpland, were in the Herbarium of M. Kunth, but the country whence they came did not appear. The plant is from a foot and a half to two feet high ; the leaves from four to seven inches long, and from one and a half to three inches broad; the flowers are bell-shaped, pendulous, three inches long, and nearly two inches in diameter, woolly outside, smooth inside. There are now twelve species of this well-limited genus, and of these one half come from the West India islands—one is from Brazil, two from Peru, one from Columbia, and two from Central America. The flowers are all more or less bell-shaped, hairy on the outside, with five distinct lobes to the corolla, yellowish or greenish, and spotted with dark red purple. The fifth pollen cell, which M. Von Martius supposes, in his description of the genus, to be abortive, is absent from all the species which I have had the opportunity of examining.—*Klotzsch, in Allgem. Gartenzeit., Jan.* 17, 1852.

531. ECHEVERIA BRACTEOSA. (*aliàs* Pachyphytum bracteosum *Link, Klotzsch, & Otto.*) A glaucous succulent undershrub. Native of Mexico. Flowers green and red. Belongs to the Order of Houseleeks. Blossoms in January and February. (Fig. 261.)

This very fine species was sent to the Royal Botanic Garden, Berlin, in 1838, from Mexico, by Mr. Charles Ehrenberg ; but we have not remarked it in English collections. Dr. Klotzsch, in publishing it in his *Abbildungen,* compared it with the genera Cotyledon and Pistorinia, from which it is very different, and overlooked that of Echeveria, forming it into a new genus, which he called Pachyphytum. It is, in fact, nothing whatever more than an Echeveria with a large fleshy calyx. The whole plant is covered with a thick glaucous bloom. The leaves grow in rosettes at the end of a short fleshy stem, are flat, obovate, obtuse, almost a quarter of an inch thick. From amongst them rises a slender leafless peduncle, clothed with narrow spathulate deciduous fleshy scales, and bearing at the end a recurved, one-sided, close raceme. The sepals are oblong, erect, united at the base into a short cup, rather unequal, and considerably longer than the dull red petals. It probably exists in our gardens among the Mexican Echeverias that have not yet flowered.

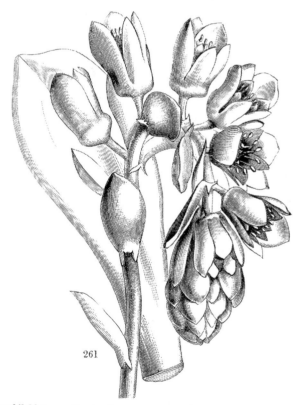

261

532. STROBILORACHIS GLABRA. *Link, Klotzsch, & Otto.* (*aliàs* Strob. prismatica *Nees; aliàs* Ruellia prismatica *Vellozo; aliàs* Harrachia macrothyrsus *Martius; aliàs* Justicia imbricata *Pohl.*) A hothouse shrub, with green cones of bracts and pale yellow flowers. Native of Brazil. Belongs to Acanthads. (Fig. 262.)

This plant has something the habit of an Aphelandra. The leaves are oblong-lanceolate, acuminate, convex, wavy bright green. The flowers are arranged in four-cornered, cones, four or five inches long, formed of strongly keeled, ovate, green bracts, from within which appear pale yellow bilabiate corollas, having a truncate two-lobed upper lip, and a three-lobed lower lip, the middle lobe of which is broader and more blunt than the laterals. The species is by no means infrequent in continental gardens, to which it was introduced from Berlin. Its native place appears to be damp shady places on the Corcovado Mountain in Brazil, and in many similar places near Rio Janeiro. We are at a loss to understand upon what principle the name first given to the plant by Dr. Klotzsch was altered by Professor Nees von

Esenbeck. It is a rule, no doubt, among some botanists to insist upon the retention of the first specific name that is published, however erroneous may have been its reference to a particular genus : the second name being held to be unchangeable whatever may happen to the first. But we dispute the propriety of this plan, and refuse to acknowledge any sufficient authority for the practice, which is sometimes impossible, very generally inconvenient, and not unfrequently absurd. Nothing is more common than for an unskilful botanist to refer a plant to a wrong genus. Another succeeds him, places it in its right genus, but with a new specific name, the first being undiscoverable on account of the original blunder with which it was associated. Then comes in a third gentleman, who takes upon himself to cancel half the first genuine name in favour of half the previous inaccurate name, and thus introduces a third name into the overburthened pages of science. For example : A publishes in 1840 a certain QUERCUS *lignea ;* B finds the plant in 1842, recognises it to be a Juglans, not a Quercus, and gives it to the world as JUGLANS *lamellata ;* then uprises C, and coolly changes B's name into JUGLANS *lignea,* upon the ground that *lignea* has a right of priority over *lamellata!* The first admissible name was in such a case *Juglans lamellata,* and to that alone, as a whole, the right of priority attaches. Naturalists cannot concede to anyone a right to interfere in the name which may be given by the first author whose entire designation is admitted to be in itself unobjectionable. For the same reason, when several new genera are founded at the expence of some old one, no one can be held to be bound to preserve all the old specific names which he may find. The new names may be wholly new, and need not be half old and half new. All naturalists of experience will preserve ancient specific names for modern genera when it is desirable, but no one can be bound to do so. It is a mistake to quote the authority of Linnæus in this matter, for his practice was precisely that for which we contend. For example : his Rheum *Rhabarbarum* had been previously called Rhabarbarum *sinense* by Ammann ; his Butomus *umbellatus* was the Juncus *floridus* of his predecessors ; his Baccharis *hali-mifolia* was the Senecio *virginianus,* &c., of Ray, and the Argyrocome *virginiana* of Petiver ; his Othonna *pectinata* was the Jacobæa *ab-sinthites* of Plukenet ; and so on in hundreds of instances. We therefore cannot acquiesce in Professor Nees von Esenbeck's change of Klotzsch's original name of Strobilorachis *glabra* into S. *prismatica,* for no better reason than that somebody (in this instance an ignorant Portuguese friar) had previously called it *Ruellia* prismatica.

533. TRIGONIDIUM RINGENS. *Lindley.* (*aliàs* Mormolyca lineolata *Fenzl.*) A dingy brown-flowered Orchid. Native of Mexico. Introduced by the Horticultural Society.

This has lately been published by Professor Fenzl, in a pamphlet called *Nova quædam genera et species plantarum,* t. 2, under the *aliàs* above quoted. It was first made known in the year 1840, in the *Botanical Register,* at No. 121 of the miscellaneous matter.

534. HAKEA MYRTOIDES. *Meisner.* A stiff-leaved greenhouse shrub. Native of Swan River. Flowers purple and yellow. Belongs to Proteads. Introduced at Kew.

Raised from seeds sent to this country by Mr. Drummond. It is extremely different from any previously described species, but perhaps most allied to *Hakea ruscifolia* La Billard. The bright red flowers (so unusual in the genus) nestled among the foliage, have a very pretty effect. It blossoms in the Royal Gardens in February. An ascendant or rather straggling shrub, a foot or a foot and a half in height, rigid, much branched; branches terete, younger ones puberulous. Leaves generally very patent and slightly tortuous, so as to have a squarrose appearance, ovate, lanceolate, sessile, subcoriaceous, with very indistinct, close-pressed pubescence, plane, or the sides slightly recurved, the margins thickened and running out at the apex into a rather long pungent mucro. Flowers in axillary, sessile fascicles, red, handsome. Pedicels purple, thickened upwards. Sepals linear, their apices spathulate, recurved, orange-yellow, bearing a yellow nearly sessile anther in the cavity. Style very long, bright red, bearing at the apex an erect, cylindrical, but rather acute stigma.—*Bot. Mag.*, t. 4643.

535. HUNTLEYA CERINA. A beautiful stove Orchid, from Central America. Flowers pale yellow, with a purple column, in April. (Fig. 263.)

H. cerina; sepalis subrotundis concavis, labello ovato convexo retuso cristâ crassâ semi-circulari truncatâ plicatâ, columnâ apice nudâ.

A third species is now added to the curious genus Huntleya, neither with brown nor violet flowers, but with firm whitish waxy blossoms, not unlike those of *Maxillaria Harrisoniæ*. It was found in Veragua, by Mr. Warczewicz, on the Chiriqui Volcano, at 8000 feet above the level of the sea, and was sold by auction by Mr. Stevens some time in 1851. Mr. Rucker has been the first to flower it. Its manner of growth and general appearance are those of *Huntleya violacea*. The flowers rise singly from the base of the leaves upon a peduncle about six inches long, with a few short tubular close-pressed scales near the base; they are very fleshy, nearly circular, concave, and about three inches across. The sepals and petals are rounded, and even at the edge, of a very pale straw-colour. The lip is somewhat ovate, convex, indented at the point, much more yellow, and furnished near the base with a deep thick semicircular ruff, composed of numerous plaits and folds. The column is deep violet near the base, and has no expansion or hood over the anther.

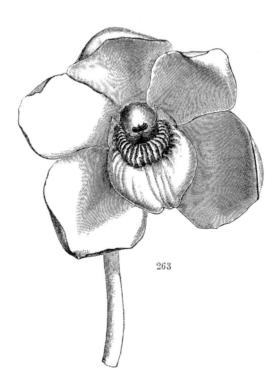

263

PLATE 85.

L.Constans del.& zinc.

Printed by C.E.Cheffins,London.

[PLATE 85.]

THE OVAL OXYLOBE.

(OXYLOBIUM OVALIFOLIUM.)

A very handsome Greenhouse Shrub, from SWAN RIVER, belonging to the LEGUMINOUS Order.

Specific Character.

THE OVAL OXYLOBE. Stipules setaceous, somewhat longer than the petiole. Leaves in whorls of three or four or opposite, oval, obtuse, or emarginate, mucronulate, silky on the under side as well as the branchlets. Heads of flowers axillary and terminal, on short stalks, densely many-flowered. Calyxes and pods shaggy.

OXYLOBIUM *OVALIFOLIUM;* stipulis setaceis petiolum subsuperantibus, foliis quaternatim et ternatim verticillatis oppositisque ovalibus obtusis v. emarginatis mucronulatis subtus ramulisque sericeis, capitulis axillaribus terminalibusque brevè pedunculatis densè multifloris, calycibus leguminibusque villosis.—*Meisner.*

Oxylobium ovalifolium : *Meisner, in Plant. Preiss.,* i. 28 ; *aliàs* Gastrolobium pyramidale : *T. Moore, in Garden Companion,* vol. i., p. 81, *with a figure.*

IT is now between twenty and thirty years since an *Oxylobium retusum,* from King George's Sound, was published in the *Botanical Register,* t. 913. The same plant had been previously described by Smith in the *Linnean Transactions,* vol. ix., p. 254, under the name of *Chorizema coriacea.* Nevertheless, the systematic writers who have followed, seem in every instance to have overlooked the plant, although it is by no means uncommon in gardens. We, therefore, reproduce the passage in which *Oxylobium retusum* was established :—

"The genus Oxylobium, as defined by Mr. Brown in the second edition of *Hortus Kewensis,* is distinguished from Chorizema of Labillardière by its calyx being nearly regular, not distinctly bilabiate; by the carina being compressed, and as long as the alæ, not inflated and shorter than alæ;

and by the pod being ovate and sharp-pointed. In the characters of the flower, the subject of this article agrees better with Oxylobium than with Chorizema, and Mr. Brown has been so kind as to inform us, that the pod is that of Oxylobium, to which genus he has referred it under the name we have adopted. A handsome greenhouse shrub, native of King George's Sound in New Holland, whence seeds were brought by Mr. J. Richardson. The specimens from which our drawing was made were communicated from Mr. Colvill's nursery, under the name of *Callistachys capitata*. Wild specimens, collected in King George's Sound by Archibald Menzies, Esq., and preserved in the Banksian Herbarium, present two forms of leaves, one ovate and the other oblong; but they are obviously only different states of the same species.

" Branches somewhat angular, furrowed, densely velvety, and ash-coloured. Stipules subulate, erect, downy. Leaves stalked, with a short, downy footstalk, oblong or ovate, retuse, with a little point, their surface elegantly reticulated with prominent veins. Racemes capitate-corymbose, stalked, axillary or terminal, much shorter than the leaves. Calyx campanulate, five-toothed, very villous, with a bractea at base, in wild specimens ferruginous, in the garden specimens silvery. Corolla orange-coloured, with purple veins. Vexillum transverse, erect, flat, emarginate. Wings and keel projecting, purple, the length of vexillum."

To this *Oxylobium retusum* the present plant is so closely allied that it is not improbable that it may be one of the forms above alluded to as existing in the Banksian Herbarium. Professor Meisner calls it *O. ovalifolium ;* and states that it was found by Preiss among close thickets near Mount Manypeak, and on rocks at the foot of the Baldhead Mountain in King George's Sound.

The main distinctions between it and *O. retusum* are that the former has the stipules much longer than the leafstalks, and the leaves as broad at one end as the other. The hairs on the shaggy calyxes are white on the stalks and tube, but rich brown on the edges and lobes, which, moreover, are very generally petaloid inside.

A very beautiful shrub, requiring the treatment applied to other New Holland leguminous plants of a similar nature. Our drawing was made from a plant belonging to Messrs. Henderson and Co., of Pine Apple Place.

PLATE 86.

L.Constans del.& zinc.

Printed by C.F. Cheffins, London.

[PLATE 86.]

THE LONG-LEAVED PUYA.

(PUYA LONGIFOLIA).

———◆———

A Stove Herbaceous Plant, supposed to come from the CARACCAS, *belonging to the Order of* BROMELIADS.

Specific Character.

THE LONG-LEAVED PUYA. A bulbous, stemless plant. Leaves of two forms ; the external spiny, leathery, narrowly pinnated, with a long awl-shaped point ; the internal grass-like, smooth, much longer than the spike. Bracts nearly smooth, shorter than the calyx. Sepals linear-lanceolate, keeled, shorter than the petals, which are rolled into a tube split on one side.

PUYA *LONGIFOLIA ;* bulbosa, acaulis, foliis biformibus, exterioribus spinosis coriaceis angustè pinnatis apice longo subulato interioribus gramineis lævibus spicâ pluries longioribus, bracteis glabriusculis calyce brevioribus, sepalis lineari-lanceolatis carinatis petalis in tubum hinc fissum convolutis duplò brevioribus.

Puya longifolia : *Morren, in Annales de la Société Royale de Gand,* vol. ii., p. 483, t. 101.

A SPECIMEN of this plant was sent to us in March last by Messrs. Weeks & Co. of the King's Road, with the flowers in the pallid state now represented. Since the plate was prepared, we have discovered that the species has been figured in the work above quoted, and that the flowers are, when in health, as deep in tint as the most scarlet Tillandsia. In Professor Morren's plant, the outer leaves were moreover broader and nearly pinnatifid, not cut down to the middle, as in ours. The account which he gives of it is this.

"This new kind of Puya possesses the coral-red brilliancy of the flowers of its congener, the *P. Altensteinii,* but its spike is much smaller. It has the habit and appearance of the *P. heterophylla* of Lindley (*Botanical Register,* 1840, t. 71), which it resembles in the bulbs, which do not flower; but in all other respects it is different. The leaves are much longer, linear, and are often as much as a foot and a half in extent, curving down around the plant, and even doubling by their own weight. The spike consists of very long straggling flowers, by no means collected into a capitate spike. The corolla of *P. heterophylla* is rose, this is as red as the richest coral; one might say that the scarlet of the bracts of *P. Altensteinii* is here transferred to the corolla, which in that species is dazzling white. The form of the nectarial scales is also different in *P. heterophylla.*"

It is uncertain when this plant came into our gardens; all that was known about it to Prof. Morren was that it was introduced into Belgium in 1843 by government collectors of plants, and he thought it highly probable that it came from Mexico. But then he adds, that it is also very probable that it came from either La Guayra or the Caraccas, where Messrs. Funck, Linden, and others had been employed.

It is strictly a stove species, demanding the treatment of Tillandsias and similar plants. It is probable that it would look best if grown like an epiphytal Orchid, which would allow the long narrow leaves to hang down without risk of being bruised or broken.

PLATE 87.

L.Constans del.& zinc.

Printed by C.F.Cheffins,London.

[PLATE 87.]

THE HOODED ONCID.

(ONCIDIUM CUCULLATUM.)

———◆———

A Stove Epiphyte, from CENTRAL AMERICA, *belonging to the Order of* ORCHIDS.

Specific Character.

THE HOODED ONCID. Pseudobulbs oval, long, bluntly ribbed. Leaves oblong-lanceolate, flat, as long as the angular scape. Raceme simple, scarcely panicled. Upper sepal and petals oval, somewhat herbaceous, equal, the lateral united into one concave oblong two-toothed body. Lip heart-shaped, fiddle-shaped, dilated at the apex, two-lobed, with round toothletted divaricating lobes ; the base furnished with three convex rounded plates, and a line of well-defined hairs near the base. Column dwarf, with short rounded auricles near the base. Anther-bed hooded, fleshy.

ONCIDIUM *CUCULLATUM,* (TETRAPETALA MICRO-PETALA) ; pseudobulbis ovalibus obtusè costatis elongatis, foliis oblongo-lanceolatis planis scapo angulato æqualibus, racemo simplici vix paniculato, bracteis parvis concavis squamæformibus, sepalo supremo petalisque ovalibus subherbaceis æqualibus lateralibus in unum oblongum concavum bidentatum connatis, labello cordato panduri-formi apice dilatato bilobo laciniis rotundatis subdentatis divaricatis lamellis 3 brevibus lævibus rotundatis pone basin serie solitariâ villosum, columnâ nanâ auriculis brevibus rotundatis juxta basin marginatâ, clinandrio carnoso cucullato.

Oncidium cucullatum : *Lindley, Sertum Orchidaceum,* sub t. 21 ; *Orchid. Linden ; aliàs* Leochilus sanguinolentus : *Bot. Reg.* 1844, misc. 91.

THIS curious plant was originally made known through a dried specimen, probably from Dr. Jameson, in Sir W. Hooker's Herbarium, gathered on the trunks of trees on the western declivity of Pichincha. It was afterwards found by Mr. Linden, in the account of whose Orchidaceous plants it is mentioned as " An epiphyte with oval obtuse ribbed pseudobulbs. This magnificent species has deep red petals, and a two-lobed violet lip spotted with purple. Forests of

Quindiu, at the height of from 7800 to 8700 feet; February. The Gallegos call it Hierba buenal and la Mesa." At a later period it was found by Schlim in New Grenada, at a place called Las Vetas, at the height of 10,000 feet above the sea.

The first knowledge we had of it in a live state was from a couple of wretched flowers sent us by the late Mr. Barker, when it was supposed to be a *Leochile*, and the following note was published of it in the Botanical Register :—

" Although the flowers are small they are very beautiful, having a deep crimson lip richly studded with clear purple spots. In the smallness of its anthers, the extension of the anther-bed behind into an elevated rim, and in the shortness of the column wings, it is somewhat different from the rest of the genus."

At last it has taken a permanent place among cultivated Orchids, and has produced the materials from which the annexed figure was made, in the collection of Thomas Brocklehurst, of Macclesfield, with whom it flowered in February last. From the gardener, Mr. Pass, we have the following note :—

" *Oncidium cucullatum* was bought at Mr. Linden's sale of imported plants, in June last. When received, it was potted in very fibrous peat and broken pots, using plenty of drainage in the pot, and placed in a rather cool and dry atmosphere, until it began to grow, when it was removed to a house used for growing Cattleyas, Odontoglots, and other South American Orchids—together with fruiting pine-plants. The heat would be from 70° at night to 85° in the day; admitting air freely on fine days, giving the plants a light syringing, throwing water on the walks, walls, &c., and closing the house early on sunny afternoons, so as to get a strong moist heat for an hour or two in hot weather. In dull cold days in summer, not uncommon here, I give air for two or three hours in the day, keeping a moist genial heat of 75° to 80° by fire. When in bloom and at rest, I keep them in a much cooler and drier house. The above is a sketch of my way of growing a portion of the Orchids here, amongst which are many of the plants sold by Linden last summer, all of which grew, and are mostly doing well. I should say that a strong-grown plant would produce more than fifteen or twenty flowers on a spike, and probably larger flowers, for our plant was very small when bought, and the bulb it made was not more than one-third the size of the imported one."

The species seems to vary a little in the colour of the flowers, which are sometimes more rose-coloured than those now represented, and in the form of the lip, a very common circumstance among alpine epiphytal Orchids.

GLEANINGS AND ORIGINAL MEMORANDA.

536. MAXILLARIA ELONGATA. A hothouse terrestrial Orchid from Central America. Flowers pale yellow and brown. Introduced by Mr. Skinner. (Fig. 264; *a*, a flower magnified.)

M. elongata, (Racemosæ); pseudobulbis cylindraceis elongatis diphyllis, foliis lanceolatis 3-costatis circiter duplò longioribus, scapo erecto bivaginato, racemo denso oblongo pseudobulbis æquali, bracteis setaceis ovario longioribus, sepalis petalisque linearibus acuminatis, labello ovato-oblongo carnosissimo utrinque lobato utrâque facie densè verrucosâ ad medium hypochilium usque.

Pale yellow flowers, as large as those of *Maxillaria supina*, with a purplish-brown lip, singularly studded within

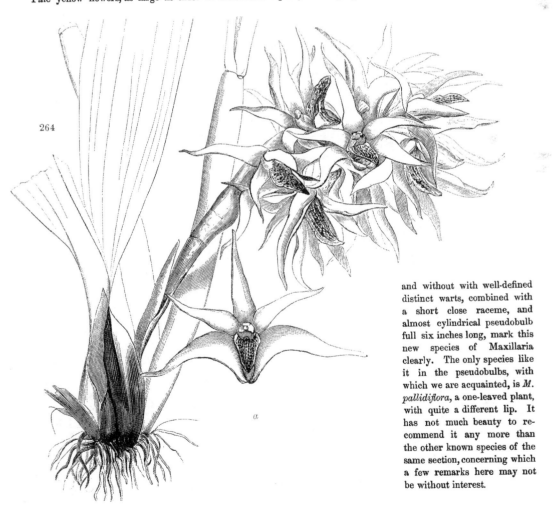

264

and without with well-defined distinct warts, combined with a short close raceme, and almost cylindrical pseudobulb full six inches long, mark this new species of Maxillaria clearly. The only species like it in the pseudobulbs, with which we are acquainted, is *M. pallidiflora*, a one-leaved plant, with quite a different lip. It has not much beauty to recommend it any more than the other known species of the same section, concerning which a few remarks here may not be without interest.

In the great genus Maxillaria, as now limited, the majority of the species (1. § ACAULES) are stemless, and produce one or two flowers only at the end of the scape, others (2. § RACEMOSÆ) are also stemless, but form their flowers in many-flowered racemes, while the remainder (3. § CAULESCENTES) have branching stems with pseudobulbs, and leaves clustered upon them at short intervals. It is to the second of these sections that the plant before us belongs, together with about a score other species, of which the following is a brief enumeration :—

2. § RACEMOSÆ (*Xylobia*).

1. M. Colleyi. *Bateman.* A brown-flowered plant, with few-flowered nearly sessile racemes.
2. M. squalens. *Hooker.* (*aliàs* Xylobium squalens *Lindley ; aliàs* Maxillaria supina *Pœppig & Endlicher.*) Flowers yellow and dirty brown, with a purple pointed lip.
3. M. scabrilinguis. *Lindley.* (*aliàs* Cyrtopera scabrilinguis *id.; aliàs* Dendrobium carnosum *Presl., Reliq. Hœnkeanœ.*) Flowers dull purplish-yellow.
4. M. bractescens. *Lindley.* Flowers dull yellow, in a tall lax raceme.
5. M. corrugata. *Lindley.* Flowers pale brownish-purple.
‡ 6. M. longifolia. *Lindley.* (*aliàs* Dendrobium longifolium *H. B. K.*) Flowers with a purple lip, on a scape two feet high.
7. M. elongata. *Of this place.*
8. M. pallidiflora. *Hooker.* Flowers greenish, in a thin raceme.
9. M. decolor. *Lindley.* (*aliàs* M. palmifolia *Lindley ; aliàs* Dendrobium palmifolium *Swartz.*) Flowers yellowish-white, in a short raceme. There can be little doubt that the two plants now brought together are identical.
10. M. concava. *Lindley.* Flowers pale yellow, in a rather thin raceme.
11. M. foveata. *Lindley.* Much like the last.
‡ 12. M. hyacinthina. *Reichenb. fil.* Flowers white, with a rose-coloured lip, very sweet-scented like a Hyacinth.
‡ 13. M. latifolia. *Lindley.* (*aliàs* Dendrobium latifolium *H. B. K.*) Flowers yellow and red, in a spike two feet long.
‡ 14. M. maculata. *Lindley.* (*aliàs* Dendrobium maculatum *H. B. K.*) Flowers large, green, very sweet-scented, in a spike rising higher than the leaves.

N.B. Those marked ‡ are not yet known to be in cultivation. The following Maxillarias, of the Flora Peruviana, are so slightly characterised, that they must all be regarded as doubtful species, which it may perhaps never be possible to identify.

15, undulata ; 16, variegata ; 17, triphylla ; 18, hastata ; 19, cuneiformis ; 20, bicolor ; and 21, tricolor.

537. PHRYNIUM SANGUINEUM. *Hooker.* (*aliàs* Maranta sanguinea *Hortul.*) A handsome stove herbaceous plant, with white flowers and crimson bracts. Blossoms in the spring. Native of — (?). Introduced by Mr. Jackson, of the Kingston Nursery.

Received from the continent, under the name of *Maranta sanguinea ;* but of what author, or where, if anywhere, it is published or described, I have not been able to ascertain. It is assuredly rather a *Phrynium* than a *Maranta*, and not very distantly removed from our *Phrynium capitatum*, figured in the *Botanical Magazine*. That species, however, differs in the colour of the flowers and the leaves, in the inflorescence, and materially in the shape of the blossoms. The plant is handsome in its flowerless state, from the rich blood-colour of the long sheaths of the petioles, and the deep purple of the underside of the leaves ; but the inflorescence adds greatly to the beauty, the upper part of the long peduncle, the copious bracts, and the flowers and pedicels and rachis being alike of a rather bright red colour. It blossoms copiously in the winter and spring months. Stemless or caulescent. Leaves ten inches to a foot long, oblong, acuminate, penninerved ; nerves oblique, dark full green above, rich purple below, on short petioles, which are jointed upon the long base, of which the inside forms a projecting membranous sheath to the scapes. Scape elongated, a foot to a foot and a half long, erect, terete, red upwards, terminated by a compound raceme, or rather compact panicle, of bracteated flowers. Bracteas all red, primary ones (at the base of the main ramifications) large, ovate, acute, conduplicate ; lesser ones, or bracteoles, of the same form and colour. Rachis short, and pedicels articulated, red. Flowers red. Ovary small, turbinate, longitudinally furrowed. The outer sepals broad, ovate, obtuse, nearly equal, free to the base, erect. Inner sepals erect, very unequal, one of them deeply two-lobed, combined for a good part of their length from below among themselves, and with the broad and flat petaloid filament and style. Anther solitary, lateral. Style curved. Stigma grooved.—*Bot. Mag.*, t. 4646.

538. CESTRUM WARCZEWICZII. *Klotzsch.* A greenhouse shrub, with light orange-yellow flowers. Belongs to Nightshades. Native of Central America. Introduced by M. Von Warczewicz.

C. glabrum ; foliis petiolatis, ovalibus, utrinque attenuatis, supra nitidis, saturate-subtus pallide-viridibus ; floribus in corymbos thyrsoideos fastigiatos terminales sessilibus ; bracteis persistentibus ; calycibus tubulosis, quinquecostatis, quinquedentatis, dentibus subulatis ; corollis infundibuliformibus, glabris, tubo flavido calyci duplo longiore, limbo vitellino, reflexo ; filamentis paullo supra medium tubum corollæ insertis, dente puberulo instructis, in lineam subpuberulam decurrentibus ; stigmate viridi, capitato, subumbilicato.

This Cestrum, which was discovered by M. Von Warczewicz near the volcano of Carthago, in Central America, was named by him in his seed catalogue *Habrothamnus aureus*; it is now (November, 1851) in flower in M. Mathieu's garden, in Berlin. It is distinguished from *C. aurantiacum* Lindl., to which it is closely allied, by its brighter green foliage and deep yellow flowers. Its botanical differences consist in its elliptical leaves, shining on their upper surface, and tapering to each end ; in its calyx, which is half as long as the tube of the corolla, or more, and has awl-shaped teeth ; and in its persistent bracts surrounding the sessile flowers. On the other hand, the leaves of *Cestrum aurantiacum* Lindl. are ovate, of a dull green ; the bracts are smaller ; the calyx is two or three times shorter than the tube of the corolla, and the flowers are of a clearer and more golden-yellow colour.—*Klotzsch, in Allgem. Gartenzeit., Nov. 15, 1851.*

539. PASSIFLORA ALBA. *Link & Otto.* A stove climber, with white flowers. Native of Brazil. Blossoms freely from May to September, and produces an abundance of fruit the size of a Walnut. (Fig. 265.)

Stem twining like that of other Passion-flowers. Leaves smooth, three-lobed, heart-shaped at the base, five-nerved, with oval lobes having mere glandular serratures at the base ; a pair of glands grows on the middle of their stalk. The stipules are cordate and half stem-clasping. The flowers, as well as their long thready coronet, are pure white, green externally ; they grow singly, with three cordate bracts at their base. This is near *P. Raddiana* of De Candolle, but the flower-stalks are not four times as long as the leaf-stalks.—*Link & Otto.*

540. TROPÆOLUM DIGITATUM. *Karsten.* A handsome annual (?) climber, from the Caraccas. Flowers bright scarlet. Introduced by M. Decker of Jena.

T. scandens, radice fibrosa, foliis peltatis quinque—septem-lobatis, lobis rotundatis integerrimisque, petalis dentato-ciliatis calycem subæquantibus et aureis, sepalis basi appendiculatis, antheris virescentibus.

The seeds of this new Tropæolum were sent by Dr. Karsten during the present year (1851), to M. Decker, who sowed them on the 8th of August, directly after their arrival. A specimen is now in flower in my garden, and the plant will be ready for sale next spring. The fibrous root of this new and beautiful climbing plant soon sends out a high climbing stem adorned with an elegant and rich foliage. The present species differs from those hitherto known in its unexampled rapidity of growth, and in the peculiar form of its leaves. The leaves are five to seven-lobed, rounded and entire, varying occasionally with respect to the depth of their lobes, of a fresh green colour, which is deepened by the greyness of the underside. From amongst this foliage the numerous yellow and carmine flowers peep out. The flower itself measures, with the spur, about one and a half inch in length. The calyx and spur are brick-red, inclining to carmine and running into pale green, the former at its base, and the latter at its point.—*Maurer, in Allgem. Gartenzeit., Dec.* 13, 1851.

541. BESCHORNERIA YUCCOIDES. A very fine half-hardy perennial from Mexico. Flowers green, among deep red bracts. Belongs to Amaryllids.

B. Yuccoides; foliis radicalibus crassis rigidis lato-lanceolatis acuminatis suprà lævissimis subtus tactu scabris margine minutissimè cartilagineo-serrulatis, scapo racemoso subpaniculato, bracteis amplis coloratis maculatis, floribus glabris tripolli caribus pedunculatis fasciculatis.

The original species of this genus, *Beschorneria tubiflora*, has no distinct stem, but produces its erect scape from the midst of a tuft of linear radical leaves, which taper into a long fine point, and are rough at the edges with very minute toothings ; they are from fifteen to eighteen inches long, by from four to six lines wide, stiff and dark green. This we learn from Kunth. In the species now published, the leaves are broad and thick, like those of *Yucca aloifolia*. The scape rises gracefully to the height of six or seven feet, with a few lateral branches ; it is smooth, blood-red, obtusely angular, and clothed at every internode with large membranous ovate crimson bracts. The flowers grow in fascicles of from two to four each, on pedicels from half an inch to an inch long, from which they very readily disarticulate ; when full-grown they are two and a half inches long above the articulation. The ovary is clavate, acutely triangular, three-celled, with numerous horizontal ovules in a double line. The sepals and petals are green, distinct, but formed into a tube, and nearly alike in form and texture, narrowly oblong, channelled, obtuse, with a thick rib at the back ; the former are more channelled and narrower than the latter ; both are yellow at the point, and become ruddy at the back ; honey is secreted in abundance from near the base, when the flowers are open ; but they never spread much at the end. The stamens are six, equal, inserted into the base of the sepals and petals ; the filaments are quite straight, and awl-shaped at first ; after a time they acquire a sigmoid form near the base in consequence of not being able to extricate themselves from the flower as they lengthen. The anthers are versatile, linear, two-celled, arrow-headed at the base, and contain a pale greenish pollen ; the pollen-grains usually adhere in fours, or a smaller number, are smooth, spherical, and have a distinctly pitted surface ; placed in water they quickly burst their outer shell, when the inner sac will escape in the form of a free transparent globe. The style is continuous with the free triangular apex of the ovary, is slender, three-cornered, and terminates in a papillose three-lobed stigma, from which drops of honey exude some time before the flower expands.

The scape of this plant contains a great quantity of singularly tough woody tubes and spiral vessels, lying in the midst of very firm colourless transparent cells. The sides of the cells, *and of the woody tubes also*, are very conspicuously marked with short oblong bars or roundish specks upon the inside of their walls. In the presence of iodine the tissue becomes pale yellow, but the bars and specks undergo no change ; they are, therefore, not protoplasm ; are they deposits of siliceous matter ?

The three genera, Agave, Furcræa, and Beschorneria, are nearly related but satisfactorily distinguished. In Agave the filaments are folded down before expansion ; in the other two they are straight. Then Furcræa has short filaments, with a great dilated base ; while in Beschorneria the stamens are long, and taper gradually from base to apex.

The plant before us flowered the other day at Abbotsbury, in the garden of the Honourable W. F. Strangways.

542. ILEX LEPTACANTHA. A handsome, hardy, evergreen shrub, from the North of China. Introduced by Mr. Fortune.

I. leptacantha ; foliis ovali-oblongis acuminatis breviter petiolatis æqualiter spinoso-dentatis dentibus gracilibus.

That this plant is an Ilex seems to be proved, in the absence of flowers and fruit, by its being readily grafted upon the common Holly. It has very handsome foliage ; the leaves being six inches long by two inches wide, of a very uniform oval figure, bordered regularly with distant slender spiny teeth. It is a good deal like the Nepal *I. dipyrena*, but that plant seems to have much more coriaceous leaves; in this plant they are of about the texture of a Portugal Laurel.

543. MEDINILLA SIEBOLDIANA. *Planchon.* A beautiful stove shrub, with rose-coloured flowers. Belongs to Melastomads. Native of the Eastern Archipelago. Introduced by M. Van Houtte.

A native, it is said, of the Moluccas, whence it appears to have been introduced to the Belgian gardens by M. Van Houtte, and through that channel to our stoves in England. It forms a handsome shrub, with large dark green leaves, and drooping racemes, of waxy rose-coloured flowers, having dark purple anthers. Our increased intercourse with the Malay Archipelago has been the means of adding considerably to our knowledge of the species of this fine genus. Twenty-four species are enumerated in Walper's *Repertorium*, and eleven additional ones are given in the *Annales* of the same author—thirty-five in all. Most of them are described in Blume's *Mus. Bot. Lugd. Bat.*, a work of great value to the student of the botany of the Dutch possessions in the Malay Islands. With us this species flowers in the spring, and continues long in blossom. Our plant is between three and four feet high, shrubby, with the stem and opposite branches quite terete ; the branchlets only are here and there seen to have an indistinct angle. At the nodes of the stem and branches, between the petioles of the leaves, is a dense tuft of soft spicules of a dirty brown colour. Leaves, on short thick petioles, four to five or six inches long, coriaceous, glabrous, between ovate and elliptical, quite entire, acute at the base, shortly and suddenly acuminate at the apex, strongly five-nerved ; nerves very prominent beneath, where the colour is pale green, while it is dark green above. Peduncle terete, as long as the finger, and, together with the thyrsoid panicle of flowers, drooping. Pedicels about as long as the calyx, which latter has the tube nearly globose, fleshy, pale rose ; the very short margin or limb erect and erose. Petals four, spreading, broad, ovate, acute, rose-coloured. Stamens eight, pointing and spreading to one side : filaments subulate, white, curved : anthers also subulate, deep purple, wrinkled on the upper side : at the base above formed into two incurved lobes, below furnished with a straight spur. Ovary combined with the calyx : style curved, subulate ; stigma obtuse.—*Bot. Mag.*, t. 4650.

544. THYRSACANTHUS RUTILANS. *Planchon & Linden.* A beautiful hothouse shrub, from Central America. Flowers rich crimson. Belongs to Acanthads. Introduced by Mr. Linden. (Fig. 266.)

266

We are only acquainted with this from the following note, and a coloured figure circulated by Mr. Linden in the beginning of the present year, of which the annexed woodcut is a copy. It has a purple round stem ; rich deep green rather undulated leaves, and drooping racemes of brilliant crimson tubular blossoms about two inches long. It seems well worth the attention of those who care for hothouse plants.

"Thyrsacanthus rutilans *Planch. & Lind.*; T. (sectionis primæ *Nees*) ; foliis subsessilibus oblongo-lanceolatis acuminatis acutis basi angustatis margine obsolete eroso-denticulatis, supra saturate-viridibus, subtus pallidis utrinque sparsim pilosulis, racemis axillaribus laxè plurifloris nutantibus ; bracteis parvis inferioribus lineari-lanceolatis superioribus subulatis, floribus ad axillas bractearum solitariis pedicellatis (pedicellis 3—4 lin. longis) calycis 5-partiti sicut rachidis crispulo-pilosuli, laciniis subæqualibus subulatis pedicellum æquantibus, corollæ tubuloso-ventricosæ subregularis coccineæ inferne sensim attenuatæ aut contractæ limbo 5-lobo. lobis subæqualibus erosis, staminibus inclusis glaberrimis sterilibus 2 brevibus capitellatis.—Cette espèce a été découverte par M, Schlim, voyageur de mon établissement, dans les endroits humides et ombragés près de Sa Cruz (province d'Ocana, Nouvelle Grenada) à un altitude de 4000 pieds. J'en ai recu des pieds vivans au mois de Juin 1851, dont quelques uns fleurissent depuis le commencement de Février et paraissent devoir durer jusqu'en Juin."

545. MASDEVALLIA WAGENERIANA. *Linden.* A curious little Orchidaceous epiphyte, from Central America. Flower small, dull pale red. Introduced by Mr. Linden. (Fig. 267.)

M. Wageneriana; uniflora, folio obovato-oblongo rotundato in petiolum angustato, scapo foliis æquali angulato, sepalis ovatis erectis æqualibus in setam longam extensis, petalis truncatis subcarnosis obtusè tridentatis margine anteriore in plicam producto, labello rhombeo serrulato apice calloso inflexo.

267

Masdevallias are among the most curious plants of their order, and sometimes among the handsomest. One of them, *M. coccinea*, which was sold lately at one of Mr. Stevens's sales, has large flowers as scarlet as a soldier's jacket. The majority, however, among which this stands, are as insignificant in appearance as they are singular in structure. Here the three sepals join into a cup, and each extends into a long flexible bristle; within the cup thus formed lie the smallest of organs of fructification, consisting of two minute truncated petals, whose fleshy front edge is folded into a kind of elbow, and whose lip is a thin lozenge-shaped serrulate plate, the end of which is callous and hooked inwards. We are indebted to Mr. Linden for our knowledge of the plant, a living specimen having been received from him in April last. Like the rest of its genus, it is a little alpine thing, requiring the treatment of a Stelis.

546. NYMPHÆA GIGANTEA. *Hooker.* A magnificent aquatic, with blue flowers. Native of New Holland. Not introduced.

During the early part of the present year (1852) seeds of an Australian Nymphæaceous plant were in the hands of several cultivators in this country as a new Victoria, *Victoria Fitzroyana*, with flowers of a "purplish-blue," from what source obtained I have not been able to ascertain. Those which were obligingly presented to us by Mr. Carter and Mr. Stokes under that name were, we think, not the seeds of a Victoria, but of a Nymphæa, and were so crushed in a letter, and sent dry, that we have no hope of their germinating. Now it does happen that we received during the past year specimens of a magnificent new *Nymphæa* from our friend Mr. Bidwill, gathered in the Wide-Bay district, North-eastern Australia, some of whose flowers certainly vie with the ordinary ones of *Victoria regia*, being a foot in diameter, and if not of a purplish-blue colour, yet blue,—the blue, as it would appear, of the well-known *Nymphæa cærulea.* We are much disposed to think that this is the plant producing the seeds in question, and that the plant having been known to other colonists in Australia, the seeds have been by them sent to their friends in this country, under the name of *Victoria Fitzroyana.* Mr. Bidwill is too good a botanist to have done so. Be that as it may, we deem it a matter of duty now to lay a figure and description of our magnificent plant before the public, and even a coloured figure; for so beautifully are the specimens dried by our valued friend and correspondent, that we think we cannot err much on that point. And sure we are that, even should all the seeds above alluded to fail to germinate, or prove to be those of another plant, our *Nymphæa gigantea* will ere long find its way into our tropical tanks, and adorn them with a Water-Lily little inferior to the royal Victoria in the size or beauty of its flowers, and with leaves equally remarkable in size, for a true Nymphæa, being eighteen inches to two feet across. A tuber which we have lately received from Mr. Bidwill for cultivation, but unfortunately dry and dead, is about the size of an ordinary apricot, and nearly as globose, having numerous depressions or eyes, like the "eyes" of the potato, with a scale at each depression. The leaves of our dried specimens are eighteen inches across, nearly orbicular, but longer than broad, with a deep fissure at the base, the margin remotely toothed, the substance very thick, and when dry coriaceous; the upper surface green, rather obscurely reticulated, the whole surface appearing minutely dotted with raised points: beneath purplish; the principal veins, very thick and prominent, radiate from the insertion of the petiole, and form elevated irregular hexagonal reticulations all over the under surface, which surface is everywhere minutely wrinkled. Petiole nearly an inch across, terete, full of air-cells; its attachment to the leaf is within, or at a distance from, the base of the fissure, and thus constitutes a peltate leaf. Flower twelve inches in diameter (in a dried state). Calyx of four leaves, or sepals, as long as the petals, broadly ovate-oblong, obtuse, green or purplish-green; one has the two margins and another one margin petaloid. Petals blue, very numerous, spreading, the outermost the largest (a few of them herbaceous at the back down the centre), obovate-oblong, that is, broadest above the middle, striated with veins, the inner ones rather shorter than the outer, linear-lanceolate, all of them obtuse. Stamens exceedingly numerous, more so than I have seen in any Nymphæaceous plant, forming a dense mass around and over the stigma; filaments filiform, short, incurved (none of them petaloid); anthers all perfect, linear, yellow, singularly curved, falcate; those in the centre obtuse; outer ones apiculate by a slight prolongation of the connectivum. Stigma so covered by the copious stamens that the structure cannot be seen without destroying the specimen.—Enough is here shown in proof that the species is very distinct from any of the hitherto blue Water-Lilies, or of the genus.—*Bot. Mag.*, t. 4647.

547. LONICERA FRAGRANTISSIMA A sub-evergreen hardy shrub. Flowers whitish, very sweet-scented. Native of China. Belongs to Caprifoils· Introduced by the Horticultural Society. (Fig. 268.)

L. fragrantissima (CHAMÆCERASUS); glaberrima, foliis sempervirentibus oblongis acutis subtus pallidis, pedunculo nutante petiolo longiore, bracteis herbaceis lineari-lanceolatis ovario longioribus.

This is one of the plants obtained from China by Mr. Fortune, while in the service of the Horticultural Society, but

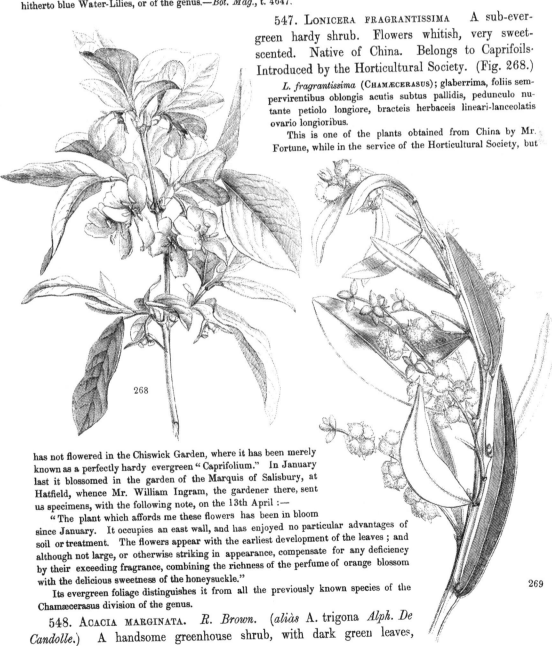

268

269

has not flowered in the Chiswick Garden, where it has been merely known as a perfectly hardy evergreen "Caprifolium." In January last it blossomed in the garden of the Marquis of Salisbury, at Hatfield, whence Mr. William Ingram, the gardener there, sent us specimens, with the following note, on the 13th April:—

"The plant which affords me these flowers has been in bloom since January. It occupies an east wall, and has enjoyed no particular advantages of soil or treatment. The flowers appear with the earliest development of the leaves; and although not large, or otherwise striking in appearance, compensate for any deficiency by their exceeding fragrance, combining the richness of the perfume of orange blossom with the delicious sweetness of the honeysuckle."

Its evergreen foliage distinguishes it from all the previously known species of the Chamæcerasus division of the genus.

548. ACACIA MARGINATA. *R. Brown.* (*aliàs* A. trigona *Alph. De Candolle.*) A handsome greenhouse shrub, with dark green leaves,

and bright yellow blossoms appearing in April. Native of King George's Sound. (Fig. 269.)

This is known in Gardens as *A. celastrifolia major*, under which name the plant from which our drawing was made was exhibited by Messrs. Henderson & Co., of Pine Apple Place. Its long narrow curved phyllodes (leaves) shorter spikes, and downy ovary, amply distinguish it from that species. To *A. myrtifolia* it approaches much more nearly, as Mr. Bentham has remarked; it seems indeed to be distinguishable only by its longer and more falcate leaves and more downy ovary. As to the *A. marginata* of Gardens, we believe it is more frequently *A. celastrifolia* itself than anything else.

549. GASTROLOBIUM VELUTINUM. A handsome Swan River greenhouse shrub, of the Leguminous Order. Flowers rich orange. Introduced by Messrs. I. and A. Henderson. (Fig. 270.)

G. velutinum; cinereo-velutinum, foliis ternis subsessilibus cuneato-oblongis v. subbilobis mucronulo interjecto margine recurvis subcrenulatis, racemis elongatis terminalibus, calycis villosi labio superiore rotundato recto bilobo inferiore 3-fido revoluto, ovario villoso stipitato dispermo.

This very pretty shrub was exhibited at a meeting of the Horticultural Society, on the 20th April last, by Messrs. Henderson, of Pine Apple Place, as a plant lately raised from Swan River seeds received from Mr. Drummond. It has in flower something the aspect of *Chorizema Henchmanni*, on account of its peculiarly rich orange-coloured flowers; but it is in reality nearer *Gastrolobium bilobum* than anything else. Its very small leaves, and soft velvety surface, are striking peculiarities.

270

550. LOASA BICOLOR. *Klotzsch.* An annual, with white flowers. Native of Central America. Belongs to Loasads. Introduced by M. Von Warczewicz.

L. herbacea, annua, robusta, ramosa, hispida, erecta; foliis alternis, petiolatis, impari-bipinnatisectis, pinnis trijugis, ovalibus, ultimis confluentibus, supra læte viridibus, sparsim-subtus subalbidis in nervis hispidis, segmentis serratis; floribus paucis, racemosis, terminalibus: calycis tubo campanulato, hispidissimo, lobis ovatis, puberulis, margine subhispidis; petalis albis, pubescentibus, calyce longioribus, apice attenuatis, setis 2 erectis, terminatis; squamis cymbæformibus, albidis, transversim coccineo-striatis.

An annual plant, one foot and a half high. It was discovered in the Chiriqui Mountains, in Central America, by M. Von Warczewicz, who marked it in his catalogue of seeds as a species of Loasa. It is at present (November, 1851,) in the gardens of Messrs. Moschkowitz and Siegling, in Erfurt. The species is closely allied to *L. rudis* Benth., from Santa Maria, in Guatemala, but is distinguished from it by the leaves which, in the present plant, are pinnated, whitish beneath, and six inches long.—*Klotzsch, in Allgem. Gartenzeit., Nov. 15, 1851.*

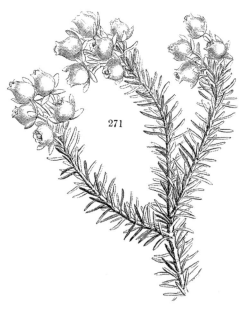

271

551. PENTAPERA SICULA. *Klotzsch.* (*aliàs* Erica sicula *Gussone.*) A half-hardy evergreen shrub, with globular pale pink flowers. Native of Sicily. Belongs to Heathworts. (Fig. 271.)

This little known plant has linear terete leaves growing in fours, globular or ovate-oblong and downy flowers, having a great spreading membranous calyx, and growing in umbel-like clusters on long slender stalks. According to Gussone the shrub grows in Sicily, on the calcareous rocks of the mountains that overlook the sea, especially on M. Cofani near Trapani. Its flowers are as large as those of an Arbutus.

552. ARAUCARIA COOKII. *R. Brown.* (*aliàs* Cupressus columnaris *Forster; aliàs* Dombeya columnaris *Forster; aliàs* Araucaria columnaris *Hooker.*)   See our Vol. II. p. 132, No. 403.   (Fig. 272.)

In the *Bot. Mag.*, t. 4635, are the following remarks upon this plant, in addition to those made in the Journal of the Horticultural Society, and quoted at the place in our work above referred to.

" To Capt. Cook, the great circumnavigator, in his second voyage, is due the first discovery of this *Araucaria*, in the little islands off New Caledonia, and subsequently on the main island :—' On one of the western small isles was an elevation like a tower ; and over a low neck of land, within the isle, were seen many other elevations resembling the masts of a fleet of ships ;' and again, a few days after, ' as we drew near Cape Coronation, we saw in a valley to the south of it a vast number of those elevated objects before mentioned, and some low land under the foreland was covered with them. We could not agree in our opinions of what they were. I supposed them to be a singular sort of trees, being too numerous to resemble anything else ; and a great deal of smoke kept rising all the day from amongst those near the Cape. Our philosophers were of opinion that this was the smoke of some internal and perpetual fire. My representing to them that there was no smoke here in the morning would have been of no avail, had not this internal fire gone out before night, and no more smoke been seen after. They were still more positive that the elevations were pillars of basaltes, like those which compose the Giant's Causeway in Ireland.' On nearing the island, a few days later, ' every one was satisfied they were trees, except our philosophers, who still maintained they were basaltes.' To the commander ' they had much the appearance of tall pines, which occasioned my giving that name to the island.' ' I was, however, determined not to leave the coast till I knew what trees these were which had been the subject of our speculation, especially as they appeared to be of a sort useful to shipping, and had not been seen anywhere but in the southern part of this land.' At length Capt. Cook landed, accompanied by the Botanists. ' We found the tall trees to be a kind of Spruce Pine, very proper for spars, of which we were in want. We were now no longer at a loss to know of what trees the natives made their canoes. On this little isle were some which measured twenty inches diameter, and between sixty and seventy feet in length, and would have done well for a foremast to the Resolution had one been wanting. Since trees of this size are to be found on so small a spot, it is reasonable to expect to find some much larger on the main and larger isles ; and if appearances did not deceive us, we can assert it. If I except New Zealand, I, at this time, knew of no island in the South Pacific Ocean where a ship could supply herself with a mast or a yard, were she ever so much distressed for want of one. My carpenter, who was a mast-maker as well as shipwright, was of opinion that these trees would make exceedingly good masts. The wood is white, close-grained, tough, and light. Turpentine had exuded out of most of the trunks, and the sun had inspissated it into a rosin, which was found sticking to them, and lying about the roots. These trees shoot out their branches like all other pines, with this difference, that the branches of these are much smaller and shorter ; so that the knots become nothing when the tree is wrought for use. I took notice that the largest of them had the smallest and shortest branches, and were crowned as it were at the top by a spreading branch like a bush ' (probably occasioned by their having been formerly densely crowded, and the tallest having most liberty at the top). ' This was what led some on board into the extravagant notion of their being basaltes : indeed, no one could think of finding such trees here.'

" There cannot be a doubt that this resemblance to columns of basalt induced the elder Forster to call this tree *Cupressus columnaris*, though he has fallen into an error in considering the Norfolk Island Pine (*Araucaria excelsa*) to be the same, as we infer from his giving ' Norfolk Island ' as a second habitat for it ; notwithstanding that Capt. Cook, in his voyage, declared it to be different. ' This ' (the Norfolk Island Pine) ' is a sort between that which grows in New Zealand, and that in New Caledonia ; the foliage differing something from both, and the wood not so heavy as the former, nor so light and close-grained as the latter.'—Of the New Caledonia Pine no perfect cones were found by the ' philosophers ' of Capt. Cook's voyage ; but a fine apex of a branch and young cone were brought home, and are preserved in the Banksian Herbarium, and figured in Mr. Lambert's splendid work, under an impression that the species was identical with that of Norfolk Island, and on the same plate with the perfect cone of the latter species. Why, under these circumstances, Mr. Lambert did not adopt Forster's name of *columnaris* we cannot conceive : we think it only justice to the latter author to restore it to that particular species for which it was intended, and to which it is so very appropriate ; we would otherwise gladly have adopted Mr. Brown's excellent one :—for assuredly nearly all the particulars we know of this interesting Pine are derived from the narrative of the illustrious navigator. Singular enough, as Dr. Lindley quotes from Mr. Moore's letter, ' the first tree of this, noticed by Capt. Cook (in 1774) as "elevated like a tower," still stands (1850) and is in a flourishing condition. Its appearance now is exactly that of a well-proportioned factory chimney of great height.' The species is no doubt equally tender with the Norfolk Island Pine."

The remarks on the nomenclature of plants made at p. 61 of the last number of this work explain why we cannot acquiesce in the name imposed upon the present Conifer by our highly valued friend Sir W. Hooker. Acting upon what we think the erroneous principle of preserving under all circumstances the specific name first given by authors to a plant, however grave may have been the errors by which that name was accompanied, our able contemporary would abolish the name of *Araucaria Cookii*, and substitute that of *A. columnaris*. Let us examine the circumstances which are said to justify this measure. The plant in question was supposed by Forster, the first botanist who saw it, to be a *Cupressus*, and he called it *columnaris*, which, had it been a Cypress, would have been a characteristic name. But it proved to have no claim to stand in the genus where it was placed, and he afterwards published it as *Dombeya columnaris*, under which name he so mixed up the present plant and the Norfolk Island Pine, that there is no certainty what he meant. When Mr. Robert Brown referred to Araucaria that plant which the late Mr. Lambert had published, in his splendid monograph of Pines, under the name of *Dombeya excelsa*, he decided, and we think rightly, that he was not called upon to go back to the name of *columnaris*, applied to Dombeya, a cancelled genus, and he preferred the well-known, though more modern, name of *excelsa*. At the same time he would seem to have been aware that Forster had confounded two different species, and to have named the new Caledonian Pine *A. Cookii*, as we learn from a statement made by the late David Don in the Linnean Transactions. That name, *A. Cookii*, was adopted in Endlicher's *Synopsis Coniferarum*, and was received in the Journal of the Horticultural Society. Nevertheless it is exchanged in the Botanical Magazine for the obsolete *columnaris*, upon the ground of posteriority of publication, although the name *columnaris* was given to a *Cupressus* or *Dombeya*, not to an *Araucaria*, although all the *Araucarias* are columnar, and the name is therefore inappropriate, and most especially although the revival of Forster's obsolete name can only tend to increase that rampant confusion among the names of plants, of which every one complains with so much truth.

The accompanying figure of the Cone is borrowed from the Journal of the Horticultural Society.

272

PLATE 88.

L.Constans del.& zinc.

Printed by C.F.Cheffins,London.

[PLATE 88.]

THE MYSORE HEXACENTRE.

(HEXACENTRIS MYSORENSIS.)

———◆———

A beautiful Stove Climber, from MYSORE, *belonging to the Natural Order of* ACANTHADS.

═══════════════

Specific Character.

THE MYSORE HEXACENTRE. Leaves oblong, acuminate, three-nerved, somewhat toothed, obtuse at the base or lobed or hastate. Bracts very small. Bractlets ovate, acute, twice as short as the corolla. Lower lip of the corolla three-parted, with ovate reflexed lobes; the upper obtuse, galeate, two-lobed; the tube at the base shaggy inside. Anthers shaggy. Stigma tubular.

HEXACENTRIS *MYSORENSIS;* foliis oblongis acuminatis trinerviis subdentatis basi obtusis lobatis hastatisque, bracteis minimis, bracteolis ovatis acutis corollâ duplò brevioribus, corollæ labio inferiore tripartito lobis equalibus reflexis superiore obtuso galeato bilobo tubo basi intus villoso, antheris villosis, stigmate tuboloso.

———

Hexacentris mysorensis : *Wight. ic. plant.*, t. 871 ; fide *Walpers' Annales*, 1. 539.

═══════════════

AMONG all the fine plants exhibited in the garden of the Horticultural Society last May, none excited such universal interest as that now represented. It formed a small umbrella-like creeper trained over trellis in the manner represented in the annexed vignette, the whole circumference of which was loaded with pendulous racemes of most beautiful large yellow and crimson flowers. The plant was sent to Messrs. Veitch of Exeter from the Mysore country, which it inhabits, as its name indicates. No doubt it is the best hothouse climber that has been introduced for many years.

We understand that the plant was sent home by Francis Maltby, Esq., of the H.E.I.C. Civil Service. Our drawing having been taken from an inferior specimen, by no means represents all the character and beauty of the species. One drawing, received from Mr. Maltby since this figure was made, represents the bunches of flowers and buds from fifteen to eighteen inches long, and

VOL. II.

M

another with the upper or first flowers dropped, and a large cluster suspended at the end of a flower-stalk of about the same length. It is added that, before the plant is out of bloom, the pendulous flower-stalks are from two to two and a half feet long.

Whatever may be thought of the so-called species, which Professor Nees von Esenbeck has separated from the original *Hexacentris coccinea*, Dr. Wallich's *Thunbergia coccinea*, nobody will question the entire novelty of the plant before us, whose small not leafy bracts, large corollas, and shaggy not smooth anthers, indicate a totally different organisation.

The genus Hexacentris, which signifies six spurs, is named in allusion to two of its stamens having one spur each proceeding from the base of the anthers, while the other two have each two spurs.

PLATE 89.

L.Constans del.& zinc.

Printed by C.F.Cheffins, London.

[PLATE 89.]

THE DWARF CRIMSON CHINESE AZALEA.

(AZALEA AMŒNA.)

———◆———

A hardy (?) *Evergreen Dwarf Shrub, from the* NORTH OF CHINA, *belonging to the Order of* HEATHWORTS.

═══════════

Specific Character.

THE DWARF CRIMSON CHINESE AZALEA. A dwarf bush. Branches when young covered with ramentaceous scales; when old rust-coloured. Leaves obovate, hairy, blunt, narrowed at the base, evergreen. Calyx wanting (?). Flowers pentandrous.

AZALEA *AMŒNA ;* humilis, ramulis ramentaceo-squamatis demum ferrugineis, foliis obovatis pilosis obtusis basi angustatis sempervirentibus, calyce nullo (?), floribus pentandris.

═══════════

THIS is a dwarf evergreen bush, resembling *Rhododendron ferrugineum* in habit. The branches when young are closely covered with long thin white ramentaceous scales; when old they are brown and coarsely hairy. The leaves are as small as those of Box, flat, obovate, very round at the point, coarsely hairy, paler on the under side. The flowers are rich crimson, almost campanulate, tolerably regularly five-lobed, with that kind of double corolla which is called "hose in hose." No calyx is discoverable; but whether that organ is absent, or is converted into the external corolla, is uncertain.

The specimen now represented was exhibited to the Horticultural Society on April the 23rd, by Messrs. Standish and Noble of Bagshot, with whom it had flowered, on which occasion it was distinguished by a Silver Knightian medal. Branches, uninjured by cold, were produced from a plant which had been exposed during the whole winter without protection; and the species is expected to be perfectly hardy. Mr. Fortune has communicated the following information concerning it :—

"This pretty Azalea was found in a nursery near Shanghae, and had been brought from the far-famed city of Soo-chow-foo. Further than this its origin is unknown. It is no doubt a very

M 2

distinct species, and probably comes from a country further north than any of its race in China, or, at all events, from a higher elevation on the mountains. As a greenhouse plant in this country it will be greatly prized. The striking form and novel colour of its flowers, its small leaves and neat habit, will render it most desirable for bouquets and for decorative purposes. But it is not unlikely that it may prove perfectly hardy in our climate; indeed it stood out in the Bagshot Nursery last winter, without the slightest protection, and flowered most profusely last spring. We may, therefore, hope to have in time a race of Chinese Azaleas growing and blooming in our borders, and vieing in beauty with the well-known Rhododendrons of North America."

Although the plant is in a monstrous state, and is clearly a garden production, yet as it seems to belong to some wild form of the genus not before described, we have felt justified in treating it as a distinct species.

PLATE 90.

L. Constans del. & zinc.

Printed by C.F. Cheffins, London.

[PLATE 90.]

THE PESCATORE ODONTOGLOT.

(ODONTOGLOSSUM PESCATOREI.)

———◆———

A Stove Epiphyte, of great beauty, from NEW GRENADA, *belonging to the Order of* ORCHIDS.

═══════════

Specific Character.

THE *PESCATORE* ODONTOGLOT. Pseudobulbs ovate, slightly ribbed, two-leaved. Leaves strap-shaped, flat, narrowed at the base, shorter than the loose many-flowered erect panicle. Bracts minute. Flowers membranous. Sepals ovate-oblong, with a small point, slightly wavy. Petals of the same form, but twice as broad. Lip heart-shaped, oblong, cuspidate, somewhat contracted in the middle, rather toothed at the base, furnished on each side with a flat lacerated appendage, a pair of parallel plates being placed between. Wings of the column short, lacerated.

ODONTOGLOSSUM *PESCATOREI* ; (Leucoglossum) pseudobulbis ovatis leviter costatis diphyllis, foliis loratis planis basi angustatis, paniculâ erectâ diffusâ multiflorâ, bracteis minutis, floribus membranaceis, sepalis ovato-oblongis apiculatis leviter undulatis, petalis conformibus duplò latioribus, labello cordato oblongo cuspidato sub-pandurato basi denticulato utrinque appendice carnoso plano lacero aucto lamellis 2 parallelis anticè denticulatis interjectis, columnæ brevis alis brevibus laceris.

Odontoglossum Pescatorei : *Linden's Catalogue.*

═══════════

NONE of the Odontoglots equal in beauty this most lovely species, to which the smallness of our plate forbids our doing justice. The panicle of large white flowers is from two to three feet high, and not much narrower, so far do the branches extend. The flowers themselves are of ample size, of a delicate semitransparent texture, with a faint blush line along the middle of the sepals, and a stain of yellow near the base of the lip, where also are found a pair of broad deep crimson lacerated appendages. The column itself is white, with the ragged wings also stained with crimson.

A specimen in flower was sent us last April by Mr. Linden, and when exhibited, although long detained on its road from Brussells, struck all who saw it with admiration. And yet Mr. Linden assures us that those very flowers had been expanded *for two months.* It had been in fact exhibited at a great Horticultural Meeting at Brussells on the 14th March, when it received a prize, which it most richly deserved. We observe that plants are offered for sale by Mr. Linden at from 100f. to 200f. each—cheap enough.

It has been named after the great and liberal French horticulturist, Mons. Pescatore, whose beautiful hothouses at Celle St. Cloud, near Paris, contain we believe the finest collection of Orchids known upon the Continent, and are perhaps richer in rare species than even the best in England.

GLEANINGS AND ORIGINAL MEMORANDA.

553. CHIONANTHUS RETUSUS.　A hardy deciduous shrub, with white sweet-scented flowers. Native of China.　Belongs to Oliveworts.　Introduced by Mr. Fortune.　(Fig. 273.)

C. retusus ; foliis longe petiolatis obovatis retusis membranaceis subtus pubescentibus, paniculis terminalibus subverticillatis nudis, corollæ tubo sepalis subulatis longiore lobis lineari-spathulatis.

Messrs. Standish and Noble of Bagshot furnished us, last May, with flowering specimens of this very pretty sweet-scented bush, obtained for them by Mr. Fortune.　When out of leaf it looks like some slender kind of Ash.　The leaves

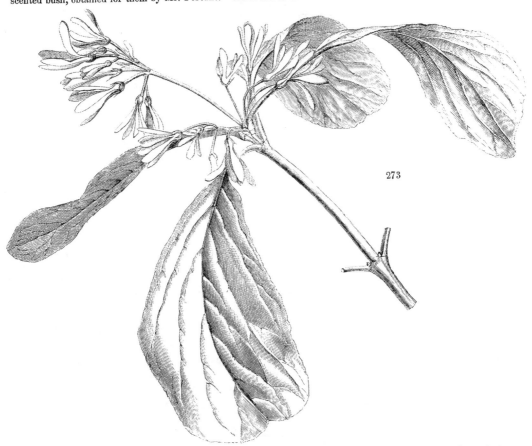

273

are slightly downy on the under side, very long-stalked, obovate, membranous, with the upper end notched out or truncate, while the lower tapers into the stalk.　The flowers are pure white, in slender, terminal, somewhat whorled panicles, shorter (in the specimens before us) than the leaves.　The corolla has a very distinct tube, rather longer than the subulate sepals, and is from four to five lobed, with the divisions linear, long, and broader at the end than at the base. There are two stamens, concealed within the tube of the corolla, with stiff, short, erect filaments, and fleshy anthers. The ovary is ovate, two-celled, with a pair of ovules in each cell. The stigma is bluntly two-lobed and sessile. We find no tendency to the separation of the petals into two parcels; on the contrary, they form a true monopetalous corolla; but they are easily torn asunder without laceration. Mr. Fortune has favoured us with the following memorandum concerning this plant :—

"A dwarf shrub, obtained in a garden near Foo-chow-foo, on the river Min. Its Chinese name is *Ting-heang*. It is deciduous in winter, and produces its snowy white flowers probably in spring. The flowers are singularly fragrant, and on this account it is much prized by the natives in Fokien. Although discovered in Foo-chow, I suspect it has been brought there from a more northern latitude. I took some plants north to Shanghae, for Mr. Beale's garden, and I observed they did not suffer in the least from a very severe winter to which they were exposed soon after they arrived. It is just possible this plant may have been brought from the Loo-choo Islands, or Japan, in the trading junks which visit Foo-chow every year. The Chinese propagate it by grafting on *Olea fragrans*. It will be better, however, to choose some other stock for it in this country, as it may probably be found a hardy plant in our climate."

554. PODOCARPUS NERIIFOLIA. *Don.* A greenhouse evergreen shrub, native of Nepal. Belongs to Conifers. Fruit fleshy, orange-red. Introduced by Dr. Wallich.

With us this forms a good-sized greenhouse shrub or small tree, with very copious dense evergreen foliage, and in a state of fruit really handsome from the copious purple-red fleshy receptacles of the seed, which are produced in the winter months. It appears to be a mountain plant, and it is not impossible but it may prove hardy enough to bear the open air, against a wall. It is with us treated like the Australian and New Zealand plants. The female flowers appear very apt to coalesce, and the receptacles then to bear two berries; and even when there is one, the receptacle seems to be often unnaturally enlarged, and to be much deformed. The male amenta are described from Dr. Wallich's dried specimens in our herbarium. The female fructification is produced in the winter months. The fleshy receptacles are said to be eaten by the Nepalese. Our plants are from six to seven feet high, much branched, the branches copiously furrowed from the decurrent petioles. Leaves scattered, approximate, sometimes appearing verticillate, in whorls of three to five, narrow, lanceolate, acute, coriaceous, the margins slightly revolute, dark green above, pale and slightly glaucous beneath, below tapering into a very short decurrent petiole. Male amenta axillary, sessile, solitary, cylindrical, slender, an inch or more long, arising from a cup-shaped scaly involucre. Anthers numerous, imbricated, two-celled, much acuminated, at length reflexed. Peduncle of the female solitary, axillary, single-flowered, about half an inch long. Receptacle of the fruit oblong, fleshy, soon enlarging, especially in breadth, with an oblong depression at the top, and variously lobed on each side, from pale yellow-green becoming orange-red, at length deep purple, slightly glaucous, bearing a small subulate recurved bractea at the base. At the apex it bears an obovate glaucous-green seed. Sometimes two or more receptacles grow from the same peduncle, and such a one we have seen to be proliferous at the extremity. —*Bot. Mag.*, t. 4655.

555. ACINETA WARCZEWITZII. *Klotzsch.* A stove epiphyte, from Central America, belonging to Orchids. Flowers pale waxy yellow, with a few red dots. Flowered in Berlin in April.

A. Warczewitzii; pseudo-bulbis ovato-oblongis, compressiusculis, leviter sulcatis, apice 3—4-foliatis; foliis maximis, late lanceolatis, subplicatis, utrinque attenuatis; scapo basilari pendulo multifloro; floribus carnosulis, pallide cerinis, apertis, perigonii foliolis exterioribus impunctatis, brevissime acutis, extus convexis, duobus inferioribus oblique ovatis, supremo elliptico, duobus interioribus æquilongis, obovatis, obtusis, basi attenuatis intusque rubro-punctatis; labello cum columna continuo, crasse carnoso, hypochilio oblongo concavo, intus puberulo, rubro-punctato, extus ad apicem umbilicato, epichilio exarticulato, tripartito, adscendente, basi appendice calloso atro-purpureo, quadrangulato, longitudinaliter unicostato, apice truncato, inflexo, lobis lateralibus latis, truncatis, erectis, intus rubropunctatis, intermedio aureo, obovato, plano, patente; columna elongata, subcurvata, albida, dorso pilosa, intus versus basin rubro-punctata, alis subangustis.—*Klotzsch.*

This is, in the opinion of Dr. Klotzsch, a well-marked new species. The scape is pendulous and many-flowered; the flowers rather fleshy, pale wax-colour, spreading open; the sepals not dotted; the petals dotted with red, as is the lip at the base; its appendage is dark purple and quadrangular; its middle lobe golden yellow. It was sent by M. Warczewicz to M. Mathieu, nurseryman, Berlin, with whom it flowered last April.—*Allgem. Gartenzeit.*, 1852, p. 145.

556. ACACIA CYCNORUM. *Bentham.* A greenhouse shrub, much like *A. pulchella.* Flowers yellow. Native of Swan River. Introduced by Messrs. Lucombe & Pince.

A. Cycnorum, as its name implies, is an inhabitant of the Swan River settlement, where it appears to be common; and Meisner gives two varieties: but Mr. Bentham is rather inclined to think that this ought to be considered, along with *A. lasiocarpa* and *A. hispidissima*, among the varieties of *A. pulchella* of Mr. Brown. Be that as it may, it is a very handsome plant, and deserves a place in every greenhouse or conservatory where early flowers are required. Shrub two to three feet high, with rather slender and scattered terete green branches, clothed with somewhat dense spreading hairs. Spines none in our specimens. Leaves alternate, bipinnate. Petiole very short, without gland (in what we have examined). Rachis hairy. Pinnæ two pairs; the lower pair each with three, the upper with four, pairs of small oblong leaflets, when dry revolute at the margin. Peduncle rather longer than the leaves, axillary, slender, arising from a scaly gemma. Head of flowers globose, rather deep yellow. Flowers crowded. Calyx turbinate, five-lobed, with spreading hairs in the upper half; the lobes short, very obtuse. Corolla four-lobed; lobes concave, ovate, erect. Stamens numerous. Style rather longer than the stamens.—*Bot. Mag.*, t. 4653.

557. SCELOCHILUS OTTONIS. *Klotzsch.* An orchidaceous, stemless, bulbless epiphyte, from the Caraccas. Flowers yellow, with a few red streaks. Introduced into the Berlin Garden. (Fig. 274.)

It is not a little singular that this rather pretty Orchid, although received in Berlin from the Caraccas in the year 1840, should never have found its way into our Gardens. In the hope of drawing some attention to it we reproduce the

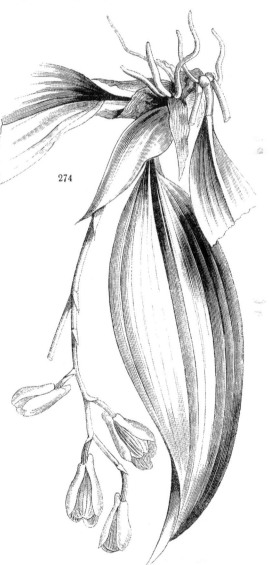

274

figure of it, from Link, Klotzsch, & Otto's *Icones*, together with a brief account of what is known about it. It was originally published in the *Allgemeine Gartenzeitung*, of August 14, 1841, with the following memorandum :—

"This small Epiphyte has, with the exception of the flowers, entirely the structure of *Oncidium carthaginense*, but the leaves are only five inches long and one inch and a half broad. The flower-spike is a little branched, slightly longer than the leaf, taper, thread-like, smooth, and covered with sessile, dry, membranaceous, lanceolate and acuminate bracts. The flowers are short-stalked, yellow, compressed, 7 lin. long. The column is without colour, twice as short as the floral envelopes. It was introduced in the year 1840 into the Botanic Garden of Berlin by Mr. Edward Otto. He discovered it upon the Silla of Caraccas, 5600 feet above the level of the sea, in thick woods, on the trunks of trees. It does not require a very high temperature, as the thermometer, at the elevation where it grows, seldom rises above $72\frac{1}{2}°$ Fahr.; it seems, likewise, to prefer the shade to the sun. It produced its small yellow flowers for the first time in the month of July."

Shortly afterwards it was republished in the work above quoted with an excellent figure, and the following amended character :—

SCELOCHILUS (Klotzsch, in Otto et Dietrich *Gartenzeitung*, 1841, p. 361.) "Perigonii conniventis foliola exteriora angusta, navicularia, carinata, basi subcohærentia, lateralia labello supposita, in unicum connata, basi in calcar obtusum, breve producta; interiora latiora, libera. Labellum integrum, supra basin columna continuum, basi brevissime bifidum, liberum, disco calloso, puberulo, longitudinaliter bicostato, antice bidentato, dentibus obtusis, conniventibus; costis infra medium bicornutis; lamina apice emarginata, subexserta. Columna semiteres, nuda, labello subduplo brevior. Anthera semibilocularis. Pollinia 2, sphærica, solida, caudicula lineari instructa, glandula parva, obovata.—Herba caracasana, epiphyta; rhizomate cæspitoso; pseudobulbis subnullis; foliis solitariis, coriaceis, carinatis, basi vaginis squamæformibus, conduplicatis, involucratis; racemo radicali; floribus compressis, flavidis.

S. *Ottonis;* foliis oblongis, coriaceis, læte-viridibus, margine acutis, subtortuosis, apice conduplicato-acutissimis, recurvis; racemo radicali subramoso, foliis parum longiore; foliolis perigonii interioribus obovatis, obtusis, intus longitudinaliter purpureo-striatis, sparsim pilosis."

The genus forms one of a small group among Vandeous Orchids, to which it has been proposed elsewhere to give the name of IONOPSIDS, and consisting of Rodriguezia or Gomeza, Scelochilus, Burlingtonia, Ionopsis, Diadenium, Comparettia and Trichocentrum. Of these *Comparettia* is known by its double-spurred lip; *Diadenium*, barbarously figured by Pöppig, although a mere puzzle, at all events must be distinct from Scelochilus; *Trichocentrum* has a long spurred auriculate lip, and distinct ecalcarate lateral sepals; *Ionopsis* has a rostrate stigma, the pollen-masses of Oncidium, and a different habit; *Burlingtonia* has a column with a pair of long arms, and the flowers of *Rodriguezia;* the latter has no spur to the sepals, ears on its column, and a free lip not rolled up in the lower sepals. It is however to

Rodriguezia that this genus comes nearest ; in fact one of the supposed species of that genus, *Rodr. stenochila* of the *Orchidaceœ Lindenianœ*, proves to be really a *Scelochilus*. A third species from Quito enables us to complete the history of the genus up to the present time, by the following enumeration :—

1. *Scelochilus Ottonis.* Klotzsch.

2. *Scelochilus Lindenii* (aliàs *Rodriguezia stenochila* of Lindley, in the *Orch. Lind.*, no. 123, where it is published with the following character:—"foliis oblongis planis, racemo laxo erecto paucifloro, floribus divaricatis, sepalo inferiore bilobo obtuse calcarato, labello angustissimo apice hastato basi sagittato, columnâ apterâ. '*An epiphyte, from the forests of Jaja. Flowers pale yellow, streaked with red. Venezuela, at the height of 6000 feet; July,* 1842.' (*No.* 659). Of this species I have only a leaf and a couple of loose scapes ; the latter are about three inches long, with two or three sharp keeled distant sheaths, and four flowers, about the size of those of *R. secunda.* The very narrow lip and thick lumpish column are quite peculiar ").

3. *Scelochilus Jamiesoni ;* foliis oblongo-lanceolatis acuminatissimis pergameneis scapo brevioribus, racemo brevi ancipiti, bracteis setaceo-acuminatis pedicellis longioribus, sepalis lateralibus semi-connatis acuminatis calcare inflato rotundato, petalis lanceolatis, labello obovato concavo apiculato basi calcare brevi didymo brachiis 2 incurvis pone basin.—*Quito. Dr. Jamieson,* 1848.—The face of the lip cannot be determined from the examination of the only flower at our disposal. It seems to be naked.

558. MAXILLARIA REVOLUTA. *Klotzsch.* A terrestrial Orchid with yellow flowers. Native country unknown. Flowered with Mr. Linau of Frankfort.

M. revoluta ; caulibus elongatis squamosis pseudobulbosis ; pseudobulbis oblongis compressis lævibus, apice unifoliatis ; folio lineari-ligulato erecto subtortuoso unicostato, apice obtuso emarginato, basi conduplicatim-attenuato ; pedunculis unifloris pseudobulbo duplo longioribus ; flore vitellino ; perigonii foliolis exterioribus oblongis erectis brevissime acutis, marginibus lateralibus brevi recurvis, interioribus brevioribus obtusis, apice revolutis ; labello erecto obtuso subtrilobo, lobo antico supra puberulo, lateralibus brevibus erectis late rotundatis, appendice linguæformi atro-purpurea adnata ad basin inter lacinias laterales ; gymnostemio erecto brevi semitereti virescente glabro, dorso obtuso ; germine longissimo tereti stricto.

This extremely pretty Maxillaria very much resembles in its habit *M. Henchmanni* Hooker, and *M. tenuifolia* Lindley, but differs from the former in its upright leaves, and from the latter in its size, and from both in the colour of its flowers, and the recurved points of its petals. The pseudobulbs are one inch long and three inches broad. The peduncle is as thick as a crowquill, upright, and provided with long, lanceolate, pointed, dry, paper-like scales. The leaves are five inches long, half an inch wide, and leathery. The ovary with its short stalk measures two inches. The sepals are from seven to eight lines long and two broad, the petals six lines long and one and a half broad ; the tongue-shaped lip five, and the column three lines long.—*Allgem. Gartenzeit.,* June 12, 1852.

559. OLEARIA GUNNIANA. *Hooker fil.* (aliàs Eurybia Gunniana *De Candolle.*) A half-hardy shrub, native of Van Diemen's Land. Flowers white. Belongs to Composites. Introduced at Kew.

This is another interesting plant of Van Diemen's Land, which braves the cold of England, and even the vicinity of London, provided it be trained against a wall. In such a position it has long been cultivated in the Royal Gardens of Kew, flowering copiously late in the autumn. We wish it had more beauty to recommend it. It was raised from seeds sent by Mr. Gunn, by whom, as its name implies, it was first detected. We think Dr. Hooker has properly referred it to *Olearia*, and that *Eurybia subrepanda*, De Cand., is merely one of the many forms of the same variable species ; variable especially in the size and incision of the leaves, and scarcely less so in the length of the peduncles and the more or less crowded flowers. Sometimes the blossoms are as copious as the leaves. A moderate-sized bushy shrub, very much branched, ultimate branches often very short. Leaves numerous, varying much in length in our native specimens, from half an inch to two inches long, on short petioles, oblong- or linear-lanceolate, generally rather deeply sinuato-dentate at the margin, penninerved, the nerves deeply impressed above, and there the surface is nearly quite glabrous, often wrinkled with reticulated veinlets : below, as on the branches, peduncles, and involucres, white with dense compact tomentum. Peduncles subterminal, on short branches, single-flowered, or elongated and panicled with several flowers or capitula, bracteolated. Involucre of several small imbricated downy scales. Florets of the ray white, of the disc yellow. Achenium, at least of the central florets, punctato-tuberculate. Bristles of the pappus rough, the scales lanceolate, with fringed serratures, sufficiently hardy to thrive in the open air of this climate in mild winters. It forms a low evergreen bushy shrub, well suited for the front row of shrubbery borders. In summer, when in flower, it presents a very showy appearance, which makes it worth while to keep a stock of young plants under protection to meet the casualties of a severe winter. It flowers freely if treated as a greenhouse plant, and is readily increased from cuttings.—*Bot. Mag.,* t. 4638.

560. LYCASTE TRICOLOR. *Klotzsch.* A terrestrial Orchid from Guatemala. Flowers pink. Introduced by Mr. Warczewicz. Flowered with Mr. Nauen of Berlin.

L. tricolor ; bracteis membranaceis elongatis acuminatis convolutis densis viridibus, suprema ovario duplo longiore ; perianthii foliolis exterioribus oblongis brevissime acutis patentibus arcuatim-recurvis pallide rufescentibus, intus ad basin subvillosis, interioribus brevioribus roseis obovatis, utrinque glabris, inferne subconniventibus, apice recurvis ; labello trilobo roseo saturate punctato glabro petalis parum breviore, laciniis lateralibus rotundatis subinvolutis, intermedia oblonga unguiculata subrecurva ; margine incisodentata, appendice ovato marginato libero suberecto brevi ad apicem inter lacinias laterales ; gymnostemio candido glabro arcuato ad basin internam purpurascente.

The pseudobulbs are enclosed in deciduous scales, long-ovate, somewhat compressed, with from six to eight blunt angles three inches long, and an inch and a half broad. They are furnished at the point with from three to five ribbed leaves, which are one foot and a half long, and from three to three and a half inches broad, longish, thin at the base, tapered into a long fine point.—The flower-stalks, of which there are generally several, seldom only one, spring from the base of the pseudobulbs, are naked, and as thick as a crowquill. The upper bract which protects and encloses the ovary is twenty-one lines long, or double the length of the ovary, ovate-lanceolate, short but finely pointed, with its edges turned towards each other on the under side, but spread out flat near the point. The sepals are long, lanceolate, sessile, very shortly pointed, of a light brown-red colour, one and a half inch long, and half an inch broad, and near the inner base a little hairy. The petals are rose-colour, obovate, fifteen lines long and seven broad. The lip is naked, rose-coloured, three-lobed towards the inside, more darkly spotted, and from thirteen to fourteen lines long ; the two side lobes are blunt, incurved, half as short as the middle lobe. The appendage, which in other species stretches from the base of the labellum to the middle lobe, and there looks like a tongue growing to its lower part, and is traceable to its base, proceeds in the present plant only from below the base of the middle lobe, and has the appearance of a flat sessile body, one and a half line in length and breadth, and whose downward course is not visible. The column is a little curved, semicircular, five lines long, white and naked.—*Allgem. Gartenzeit., June* 12, 1852.

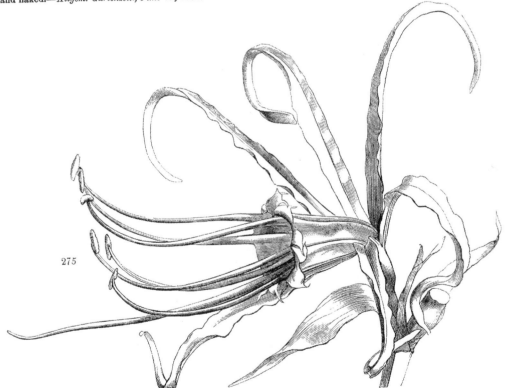

561. ELISENA LONGIPETALA. *Lindley.* A half-hardy bulbous plant. Flowers white tinged with green. Native of Peru. Belongs to Amaryllids. (Fig. 275.)

This plant was first noticed in the *Botanical Register* for 1838, p. 45 of the miscellaneous matter, with the following memorandum :—

"To the cultivators of bulbous plants this fine species will form a welcome addition. It is very nearly related to the *Pancratium ringens* of the Flora Peruviana, out of which Mr. Herbert has formed his genus Elisena, and, like it, is a native of Peru. It was obtained from Lima by Richard Harrison, Esq., of Aighburgh, near Liverpool, and it blossomed in the stove of that gentleman in May, 1838. The leaves are much like those of an Amancaes ; the flowers are of a delicate semi-transparent white, and are remarkable for their long weak sepals, which are rolled up, and in that state scarcely wider than the long white declinate stamens."

Its flowers have lately been sent us by an unknown correspondent, and have enabled us to give the following figure of one of them. About five such grow in an umbel at the end of a stiff two-edged scape, about three feet high. Dean Herbert's figure, in the *Botanical Magazine*, t. 3873, does not at all do justice to the species, which is really very handsome. He recommends it to be grown out of doors in a bed of white sand, and guarded against spring frosts.

562. BRACHYSEMA LANCEOLATUM. *Meisner.* An evergreen greenhouse shrub, with rich crimson flowers. Belongs to the Leguminous Order. Native of Swan River. Introduced by Messrs. Lucombe & Pince.

A handsome species, and its beauty is enhanced by the good-sized almost polished leaves, dark green above, beautifully silky beneath. It is a native of Swan River, and was raised from seeds sent home by Mr. Drummond, in the Exeter Nursery of Messrs. Lucombe, Pince, and Co., where it flowered for the first time in February, 1852. It is one great charm of the Australian plants that they so generally flower when there is little else to enliven the conservatory, and this cannot fail, on that account, to be very acceptable to cultivators. Dr. Meisner had evidently very imperfect specimens to describe from, for he was ignorant of the colour of the corolla, which in the living and in the dried specimens of Mr. Drummond is of the richest scarlet; and he describes the flowers as solitary. Yet he has contrived to form three varieties. The leaves are certainly variable in form, even on the same individual branch. A handsome though somewhat straggling shrub, with terete, silky branches, and usually opposite leaves, from two and a half to three inches long, shortly petiolate, varying from ovate to lanceolate, rarely obtuse, usually acute and mucronate, quite entire, penninerved, the upper surface dark green, and when dry beautifully and minutely reticulated. Petioles at most two lines long, with a subulate, coloured stipule on each side, eventually probably deciduous. Flowers four to six, on a sessile subcompound raceme in the axils of the leaves, and shorter than the leaves. Bracteas ovate, acute, silky. Pedicels short. Calyx large, ovate, five-lobed; lobes acuminate, erect. Corolla, all at least that is distinctly visible, rich scarlet; for the *alæ* and *vexillum* are scarcely protruded beyond the calyx, while the carina is twice the length of the latter. The small vexillum is cordate, attenuated, yet obtuse, white at the margin, red in the disc, with a large yellow spot in the centre. Stamens ten, free. Ovary oblong, silky. Style subulate-filiform. Stigma obtuse.—*Bot. Mag.*, t. 4652.

563. CORDYLINE INDIVISA. *Kunth.* (*aliàs* Dracæna indivisa *Forster*). A hardy (?) arborescent Yucca-like plant, native of New Zealand. Flowers in large whitish fragrant panicles. Belongs to Lilyworts. Introduced by Messrs. Veitch of Exeter.

A portion of this noble plant, consisting of a few leaves and a piece of the inflorescence, was exhibited by Messrs. Veitch of Exeter at the July meeting of the Horticultural Society, it having flowered in their nursery at Exeter for the first time in Europe. It is stated to be an inhabitant of Dusky Bay in New Zealand, where it grows as much as eighteen feet high on rocks near the sea. At Exeter it forms a noble specimen, twelve or fourteen feet high, with a single graceful stem, terminated by hard sharp-pointed sword-shaped leaves nearly four feet long by two inches wide, and narrowed into a very slender point; they are pale bright green, and perfectly smooth to the touch on both sides. From their centre springs a panicle some three or four feet long, of many compound branches, the ultimate divisions of which are graceful many-flowered angular spikes. Each flower sits in the middle of ovate scarious bracts, and consists of a short cup with a white six-parted spreading limb, of narrow blunt concave segments, at the foot of each of which is placed a stamen with a broad petaloid filament. The ovary is obovate, three-celled, with many axile ovules in each cell; the style is filiform, the stigma simple. The plant is, therefore, a *Cordyline*, and not a *Dracæna*. Nothing can be more deliciously fragrant than the flowers of this fine plant, which reminds the observer of the stately *Yucca draconis*, of which it has all the habit, but much lighter green leaves. It has lived for many years in the open ground in the Exeter Nursery, and seems to be quite hardy. According to Richard it produces blue globose berries, each marked with three excavated points near the end, and containing about seven dark smooth roundish, half-moon-shaped seeds in each cell.

564. MORMODES FLAVIDUM. *Klotzsch.* A terrestrial Orchid, with yellowish flowers. Native of Central America. Introduced by M. Von Warczewicz. Flowered with Mr. Mathieu, nurseryman, Berlin.

M. flavidum; pseudobulbis elongatis, articulatis, vaginatis, versus apicem articulatim foliosis; racemo paucifloro, pedunculato pseudobulbo altiore; floribus flavidis, erectis, pedicellatis, bracteolis oblongis, obtusis, aridis, albicantibus instructis; perigonii foliolis lanceolato-linearibus, acutis, flavidis, tribus exterioribus arcte reflexis, binis interioribus erectis; labello albido-luteo, erecto-incurvo, obovato, apiculato, integerrimo, lateribus deflexis; columna oblique torta, acuminata.

The pseudobulbs are long, cylindrical, furnished with six or seven joints, covered with sheath-like bases of leaves, and four inches long by three-quarters of an inch thick. The leaves of the specimen which I possess are not developed; the flower-stalk springs from the third joint of the pseudobulb, is as thick as a crow-quill and three inches long, but is not fully formed, for its point, near which are three empty bracts, is evidently curved, whilst below its point two normal expanded flowers appear. The bracts, which half surround the flower-stalk, are whitish, dry, oblong, rounded at the point, and three lines in length. The flower-stalks are eight lines long. The greenish-yellow sepals are from fourteen to fifteen lines long, and three lines broad towards the base, but become gradually smaller towards the top. The lip is obovate, yellowish-white (as is the acuminate column), bent inwards, with a short point, almost entire, with both edges curved back, from ten to eleven lines long, and below the point six lines broad.—*Allgem. Gartenzeit.*, *April* 10, 1852.

565. GUICHENOTIA MACRANTHA. *Turczaninow.* An inelegant greenhouse shrub, with pale purple veiny flowers. Native of Swan River. Belongs to Byttneriads. Introduced at Kew.

A singular-looking, rather than beautiful, hoary shrub, with large purplish flowers, at first sight not unlike those of some Solanum ; native of Swan River, whence seeds have been sent by Mr. Drummond to Kew, and reared in 1847. Our first flowers appeared in March, 1852, in an ordinary greenhouse. The genus Guichenotia, so named by M. Gay, in compliment to the gardener of M. Baudin's expedition, M. Antoine Guichenot, was founded upon the *G. ledifolia,* equally with this an inhabitant of the Swan River district, and is described by Mr. Turczaninow from Mr. Drummond's dried specimens. It is an extremely distinct species. The shrub is with us two and a half feet high, erect, branched. Branches terete, clothed with stellated down. Leaves downy, whorled in threes, linear-oblong, on very short petioles, entire, penninerved, the nerves almost at right angles from the costa, transverse, slightly branched, the margin revolute. Peduncles axillary, generally longer than the leaf, erect, few-flowered ; flowers one to three, drooping. Pedicels naked. or bearing one to two lanceolate distinct bracts: the hypocalycinal bract tripartite, appressed, leafy, veined. Calyx between rotate and campanulate, dull and pale purple, downy, veined, the five lobes acuminate. Petals five, small, squamiform, dark purple, one at the base of each stamen. Stamens converging into a cone against the pistil : filaments subulate : anthers dark purple. Germen ovate, acuminate, downy. Style articulated upon the ovary, about equal to it in length, slender, subulate. Stigma obtuse. —*Bot. Mag.,* t. 4651.

566. CLAYTONIA ALSINOIDES. *Sims.* (*aliàs* C. unalaschkensis *Fischer* ; *aliàs* Limnia alsinoides *Haworth* ; *aliàs* C. sibirica *Bot. Mag.,* with pink flowers.) A neat succulent annual, with small white flowers. Native of North West America. Belongs to Purslanes. (Fig. 276.)

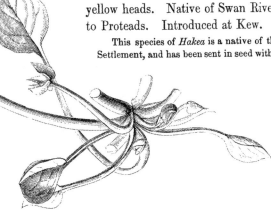

276

A small annual, with bright green succulent insipid leaves, forming patches eight or nine inches in diameter, and well suited to form a temporary covering to waste places or borders that require to be concealed without being cropped. Its flowers are white, or in the Siberian variety pink, small, but pretty when open beneath the sun. It seems to be common all over North West America, and is sometimes found apparently wild in England, that is to say in wild places to which it has been carried by birds, which eat the seeds greedily. The separation of the upper leaves at the base makes it impossible to confound it with the common *C. perfoliata,* which is, however, very nearly allied to it.

567. HAKEA SCOPARIA. *Meisner.* A long slender-leaved greenhouse shrub. Flowers in yellow heads. Native of Swan River. Belongs to Proteads. Introduced at Kew.

This species of *Hakea* is a native of the Swan River Settlement, and has been sent in seed with corresponding dried specimens (numbered 600) by Mr. Drummond. It is evidently the plant described by Dr. Meisner in the *Plantæ Preis-siance* above quoted, from specimens of Mr. Drummond in Mr. Shuttleworth's herbarium. The author, indeed, thinks it possible it may prove to be a variety of *H. sulcata*, but to us it appears unquestionably different, and the distinguishing cha-racters are well pointed out by Dr. Meisner. A small shrub, with rather tortuous terete branches, clothed with pale grey bark, the younger ones puberulous. Leaves alternate, eight to ten inches long, about as thick as a blackbird's quill, elongated, filiform, rigid, semiterete, rather deeply five-furrowed throughout their whole length, the upper furrow the broadest, hairy in the furrows, the apex sharply mucronate, the base, where inserted upon the branch, a little swollen and dilated. Flowers pale yellow, arranged in sessile heads, which are axillary, involucrate, involucre of several imbricated, brown, pubescent scales, shorter than the heads. Pedicels as long as the perianth. Perianth of four spathulate pale yellowish-white sepals, the apices concave, reflexed. Style very long, a little dilated at the apex, and there bearing a nearly cylindrical stigma.—*Bot. Mag.*, t. 4644.

568. MAXILLARIA HARRISONIÆ. *Lindley.* A stove Epiphyte from Brazil, with large waxy pale yellow flowers, and a rich rose-coloured hairy lip. Flowers in April and May. (Fig. 277.)

Of this common plant, with which all growers of Orchids are now acquainted, there are two striking varieties; one with nearly white flowers, except the lip, which is, as usual, rose coloured; the other, now figured, with smaller flowers than common, a rather shorter spur, and a much narrower lip, which has clearer veins on its lateral lobes. The specimen figured was exhibited last April by Mrs. Lawrence, and has since appeared in other collections. Of its history nothing is known. We should add that the pseudobulbs are rather narrower than in the original species, but the leaves are not at all different.

277

PLATE 91.

L.Constans del.& zinc.

Printed by C.E.Cheffins, London.

[PLATE 91.]

THE THREE-FLOWERED ABELIA.

(ABELIA TRIFLORA.)

◆

A half-hardy Shrub, from NORTHERN INDIA, *belonging to the Order of* CAPRIFOILS.

Specific Character.

THE THREE-FLOWERED ABELIA. Leaves ovate-lanceolate, entire, subsessile, ciliated. Flowers in threes; the lateral with three bracts. Calyxes shaggy, five-parted, with linear very narrow, acuminate, divisions as long as the tube of the corolla.

ABELIA *TRIFLORA ;* foliis ovato-lanceolatis integris subsessilibus ciliatis, floribus ternatis : lateralibus tribracteatis, calycibus villosis 5-partitis laciniis linearibus acuminatis angustissimis corollæ tubi longitudine.

Abelia triflora : *R. Brown, in Wallich's Plantæ Asiaticæ rariores*, vol. i., p. 14, t. 15.

FOR living specimens of this beautiful shrub we are indebted to Mr. Moore of Glasnevin, who sent them last June, with the following memorandum :—

"*Abelia triflora* is now nicely in flower here in the open border, where it has stood in front of one of our conservatories without protection, since it was planted four years ago. Major Madden sent the seeds here from Simlah, from which our plants were raised in 1847, and this is the first of them which has bloomed. I consider it an acquisition in the way of a hardy shrub. Our plant is about three feet high, and covered over with pretty pink blossoms."

Dr. Wallich states that it is found wild on the highest mountains of the province of Kamaon, towards the Himalaya, where his plant-collector, Robert Blinkworth, met with it in the month of May; the natives called it *Kumki*. He reported it to be a small tree, with delightfully fragrant blossoms, like those of *Jasminum revolutum*.

The branches are slender, grey, and covered with long hairs. The leaves are very dark green, bordered with red, ovate-lanceolate, very acute, slightly silky on both sides, and copiously fringed with long hairs. The flowers, which appear at the ends of the branches in clusters of threes, are remarkable for the very long hairs which cover the five narrow sharp-pointed reddish erect sepals, and which are as long as the tube of the corolla. The latter is pale yellow before expansion, but when open with a flat white limb, having five rounded lobes delicately tinted with rose.

This must be regarded as a charming addition to our shrubs, even although it should in England require a greenhouse or a conservative wall. In Dublin it seems to be hardy; but experience tells us that we must make some allowance for the greater mildness of an Irish winter.

PLATE 92.

L.Constans del.& zinc.

Printed by C.F.Cheffins,London.

[Plate 92.]

THE LARGE-FLOWERED GLUTINOUS DIPLACUS.

(DIPLACUS GLUTINOSUS; *var.* GRANDIFLORUS.)

◆

A Greenhouse Evergreen Shrub, from CALIFORNIA, *belonging to the Natural Order of* LINARIADS.

Specific Character.

THE GLUTINOUS DIPLACUS. Branches downy. Leaves oblong or lanceolate, rather obtuse, irregularly toothed and eroded or entire, narrow at the base, smooth on the upper side. Flowers solitary. Calyx smoothish, with lanceolate unequal teeth.

DIPLACUS *GLUTINOSUS;* ramis pubescentibus, foliis oblongis lanceolatisve obtusiusculis eroso-dentatis integerrimisque basi angustatis supra glabris, floribus solitariis, calycis glabriusculi dentibus lanceolatis inæqualibus.—*Bentham.*

Diplacus glutinosus : *Nuttall, in Taylor's Annals of Natural History,* t. 138 ; *Bentham, in De Candolle's Prodromus,* t. 368 ; *aliàs* Mimulus glutinosus *Wendland obs.* p. 51.

Mʀ. BENTHAM has truly remarked (*De Cand. Prodr.* x. 368) that this species is extremely variable ; the stem being more or less woody ; the leaves from two to six inches long, and from four to twelve lines broad, blunt or occasionally rather sharp-pointed, coarsely toothed or hardly toothed at all, more or less downy or flocculent on the under side ; the flower-stalk as long as the calyx or more usually shorter ; and the corolla of very uncertain length, pale yellow, orange, or crimson, with the lobes more or less deeply divided. These conclusions are abundantly justified by the evidence to be found in gardens, no less than by the long series of specimens in his own herbarium.

For cultivators the species may be separated into the following varieties :—

1. AURANTIACUS, figured in the *Botanical Magazine,* t. 354, with orange-coloured flowers.

2. PUNICEUS, figured in the same work, t. 3655, with scarlet flowers, the lobes of which are very shallow.

3. GRANDIFLORUS, now figured with pale salmon-coloured flowers, the lobes of which are deeply cleft, and the leaves rather shorter than usual, and less serrated.

4. LATIFOLIUS, with large yellow flowers, whose lobes are scarcely split, broader leaves, and, as is said, a dwarfer habit.

All are greenhouse shrubs, found on the borders of streams and damp situations in California, where they grow about six feet high. That now figured has been raised in many places within the last few months, and has already gained the false name of *D. leptanthus,* a plant to which it bears very little resemblance.

Flowering early in the spring, having a neat habit, succeeding well with unskilful people, and propagating freely by cuttings, the species of this genus have always been favourites, and will long continue so. That now figured is certainly much the finest, on account of its large pale salmon-coloured flowers ; a cross between which and *puniceus* ought to be very handsome.

PLATE 93.

L.Constans del.& zinc.

Printed by C.F.Cheffins,London.

[PLATE 93.]

THE FIERY-RED MORMODES.

(MORMODES IGNEUM.)

◆

A Hothouse Epiphyte, from CENTRAL AMERICA, *belonging to the Natural Order of* ORCHIDS.

Specific Character.

THE FIERY-RED MORMODES. Raceme long, many-flowered. Sepals reflexed, petals ascending: both lanceolate, flat, very acute. Lip stalked, fleshy, with a distinct point, rolled back at the sides, scarcely angular, with a transversely elliptical outline.

MORMODES *IGNEUM ;* racemo elongato multifloro, sepalis reflexis petalisque ascendentibus lanceolatis acutissimis planis, labello unguiculato carnoso apiculato lateribus revolutis ambitu transversè elliptico vix angulato.

THIS fine plant, and several others of the same genus, has been produced from the *rejectamenta* of one of Mr. Warczewicz's sales. In January last, we received from Mr. Rucker five sorts of Mormodes, all derived from the same source, all in flower, and all new. Of these we represent three.

That in the middle of our plate, to which the name of *igneum* is given, was conspicuous for the greater size of its parts, and for its intense colouring. A stiff stalk, about a foot high, bore a dozen large fleshy flowers, of which the sepals and petals were alike chocolate-coloured, and the lip a rich fiery orange-brown. There was no streaking or spotting in any part of the surface. The sepals were flat, linear-lanceolate, very sharp, and spread flat out, even turning backward after a time ; the petals, on the contrary, were erect, and somewhat broader. The lip, a tough fleshy body, when spread out had an elliptical outline, with the major axis transverse, and the edge extended into a triangular point on one side ; in its natural condition it was rolled back, and folded so as to look as if angular, though not really so.

The sorts marked B and C in the plate accompanied it. B had dingy red flowers, marked with lines of dots; and C had dark lake flowers, speckled irregularly with red, but not dotted; their lips were thinner, smaller, and had a decidedly angular outline.

The two other kinds, not now figured, were the same in habit; but neither had any dots; one had faint stripes along the sepals and petals, which were dirty pink, and the lip was a dull green; the other had a much yellower flower; in both the sepals and petals were as in B and C, but the lip was much larger, thinner, and still more decidedly angular.

Are these forms to be regarded as distinct species? and are they new, or are they varieties of some species already known? There grows in the temperate parts of the snow-capped mountain ridge of Santa Martha, especially on the branches of an Erythrina, a Mormodes of which travellers speak as being most remarkable for the infinite variety of its colours. A striped state of it having flowered at Syon, some years since, Sir William Hooker published it in the *Botanical Magazine*, t. 4214, and called it *Cartoni*, under which name it is current in gardens. Of that plant we entertain no doubt that our figures B and C are mere varieties. The main figure, so resplendent in colour and striking in dimensions, seems to differ in its broader and more fleshy sepals and petals, and in its thicker more leathery lip, which has little of the angularity which belongs to *M. Cartoni*; we therefore distinguish it under the name of *M. igneum*. As for the other varieties above alluded to, and not figured, they probably belong to the *M. flavidum* of Klotzsch.

It is not improbable, however, that all these things are one and the same species; and if so the *M. lentiginosum* of the *Botanical Magazine*, t. 4455, will have to be added; for beyond colour the plant seems to have nothing to distinguish it except the total absence of all angularity in the lip. The same principles which justify the separation of that plant equally authorize the distinction of *Cartoni*, *igneum*, and *flavidum*; and also the separation of a small species with rather more membranous pallid flowers, also from Santa Martha, and now in our gardens, the lip of which is rolled up into a slender pipe, but which when flattened has much the form of a sharp trowel. We received it last March from an anonymous correspondent at Buckland in Berkshire, and propose to distinguish it with the following name and character.

M. convolutum; sepalis petalisque linearibus reflexis, labello tereti convoluto unguiculato apiculato incurvo lævi ambitu hastato angulis abbreviatis et igitur trullæformi.—*Santa Martha.*— Flowers the smallest yet known in the genus, dull yellow, spotless.

A much more striking species than any yet recorded was sold at the same sale of Mr. Warczewicz's as the others. It formed Lots 39 the first day and 34 the second day. According to a drawing now before us, for which we are indebted to Mr. Skinner, the flower-buds are three inches long, and consequently each flower, when expanded, is eighteen inches in circumference. They are represented as of a deep chocolate-brown, and are especially remarkable for the lip being ovate-lanceolate, taper-pointed, and *perfectly flat*. It may be distinguished thus :—

M. macranthum; racemo laxo multifloro, sepalis petalisque angustè lanceolatis acuminatis patentibus, labello unguiculato ovato-lanceolato acuminato plano.—*Central America*, 7000 *feet above the level of the sea.*

GLEANINGS AND ORIGINAL MEMORANDA.

569. BURLINGTONIA DE-
CORA. *Lemaire.* (*aliàs* B.
amœna *Planchon in Hort.*)
A beautiful epiphyte from
Brazil. Flowers rose-co-
loured spotted with red, and
a white lip. Introduced
by M. de Jonghe, of Brus-
sels. (Fig. 278.)

B. decora; pseudobulbis compressis ovatis monophyllis, foliis lanceolatis subundulatis recurvantibus,
racemis laxis 3—5-floris, sepalis petalisque conniventibus acutis lateralibus vix semiconnatis, labello multò
longiore bilobo dilatato basi appendice lacero pubescente
flabellato colorato utrinque aucto, calcare conico brevi—
" staminodiis antenniformibus pilosis rubris gynostemium
æquantibus, styli cornubus glabris staminodiis plus duplo
brevioribus."—*Planchon.*

This very pretty epiphyte has been lately figured by
M. Van Houtte in his *Flore des Serres*, with a note by
M. Planchon, of which the following is the substance :—
Introduced from the province of St. Paul's, in Brazil, by
M. Libon, the collector for M. de Jonghe, this flowered
in May, 1851, with M. Makoy, when it was provisionally
named *B. decora*, under which name it is mentioned in
various trade-catalogues. It was afterwards published
as *B. decora* by M. Charles Lemaire, in the "*Jardin
Fleuriste, II., jan.* 1852, *t.* 188."

The habit is that of *B. rigida*, but the sepals and
petals are deep rose-colour, spotted with small irregular
crimson specks ; the lip, which is twice as long as the
sepals, is pure white, with a lacerated pinnate red and
speckled appendage on each side of the base. It seems
to prefer a mixture of sphagnum, rotten willow-wood,
and broken potsherds, in which it succeeds perfectly,
suspended in a basket of copper wire. It likes a hot
damp atmosphere while growing, and a good season of
rest, obtained by lowering the temperature, and diminishing the humidity of the atmosphere.

278

M. Planchon naturally compares it with a *B. obtusifolia*, very slightly defined in the *Sertum Orchidaceum*, under t. 36;
and it is indeed a member of the same division of the genus, characterised by the column bearing a pair of long hairy
ears (staminodia). But *B. obtusifolia* is in reality very near *B. rigida*, from which it only differs in the ears being
blunter and longer, the leaves smaller, narrower, more blunt, and tapering to the base, the flowers smaller, and the

lateral lobes of the lip much narrower. In this plant, however, we have, according to the authors above quoted, much smaller flowers, a simple conical not two-lobed spur, short very sharp sepals and petals, and a pair of great lacerated appendages at the base of the lip ; to say nothing of the spotting which is so much unlike anything known among Burlingtonias, except *maculata*.

570. RHODODENDRON LEPIDOTUM. *Wallich.* (*aliàs* Rhododendron elæagnoides, R. salignum, and R. obovatum *Hook. fil.*) A pretty alpine greenhouse shrub, with yellow or purple flowers. Native of the Himalayas. Introduced at Kew.

The purple-flowered state of this very variable species of Rhododendron blossomed freely in April, 1852, in a cool greenhouse of the Royal Gardens. The seeds were sent from Sikkim-Himalaya by Dr. Hooker, under the name of *R. elæagnoides*, and as such this is figured in the work on the Rhododendrons, with dark purple flowers, and also with deep yellow flowers, looking like those of some *Helianthemum*. In that work, however, the author alludes to its close affinity, as well as that of *R. salignum*, with the *R. lepidotum* of Wallich (only known to us from dried specimens) ; and a further examination has satisfied him that they and his *R. obovatum* can in no way be specifically distinguished from authentic specimens of *lepidotum*. He has, therefore, in the *Journal of the Horticultural Society of London*, united them. "The species abounds," Dr. Hooker says, "at an elevation of Eastern Himalaya of from 14,000 to 15,000 feet; but may be found as low down as 8000 feet, in moist valleys, forming a stout tortuous stalk : the branches as thick as a crow's quill, rather scattered, bearing tufts of branchlets at the top. It is a slender or stout twiggy shrub, one to four feet high, branching, often growing in widely extended clumps, as heather does with us, but never so extensively ; and it emits in sunshine a powerful resinous odour. Leaves of a pale glaucous green, lighter underneath, and sometimes ferruginous where the scales abound, one half to one and a half inch long. Flower-stalks more or less elongated, one and a half to two inches long, slender. Corolla yellow or dirty purple, half an inch across the lobes, scaly, especially on the outside of the tube ; the upper lobes are spotted with green. The odour of this plant is strongly resinous, and rather sweetish and pleasant. Its common native name is *Tsaluma*, or *Tsuma*, amongst the Bhoteas."—*Bot. Mag.*, t. 4657.

571. VERONICA ELLIPTICA. *Forster.* (*aliàs* V. decussata *Aiton.*) A hardy (?) evergreen bush, with deep green leaves and white flowers. Native of the antarctic and neighbouring regions. Belongs to Linariads. (Fig. 279.)

Beautiful flowering specimens of this were exhibited last spring to the Horticultural Society, by the Hon. W. F. Strangways, with whom the plant is hardy in Dorsetshire. It forms a dwarf dark green bush, with opposite oblong leaves, each pair of which regularly crosses the previous pair, so as to produce the appearance which botanists call decussate, the name by which the plant is known in gardens. Dr. Hooker has, however, ascertained that in reality it is the same plant as the *V. elliptica* of Forster, published many years before the name *decussata* was heard of. Upon what ground this opinion has been formed will appear from the following extract from Dr. Hooker's excellent *Flora antarctica*, vol. i., p. 58 :—

" Found in Lord Auckland's group and Campbell's Island ; at the margins of woods near the sea, abundant.

" This is a very well known plant in our gardens, introduced from the Falkland Islands, and is one of the most antarctic trees, both in this longitude and in that of extreme Southern America, there reaching the fifty-seventh parallel of latitude. It was first collected in New Zealand by Forster, its original discoverer, in Dusky Bay, where it has since been found by Anderson and Menzies. I believe it, however, to have been noticed before as a native of the Straits of Magalhaens, by the older navigators.

" In combining the *V. decussata* Ait. with *V. elliptica*, I have followed the unpublished opinion of Dr. Solander. In the British Museum there are drawings of the latter plant by Forster, New Zealand specimens collected probably by that author, and notes by Dr. Solander. The specimens alluded to are in fruit only, and agree in the foliage with the figures, which represent it in its flowering state. Dr. Forster's own handwriting (of *V. elliptica*) is on the same sheet with it ; but another plant, *V. Menziesii* Benth. MSS., has been fastened down on the paper at a future period, and the habitat, 'New Zealand, Dusky Bay, Gul. Anderson,' is written on the back, a station probably applying to the latter specimen alone. Solander's handwriting of *V. decussata β.*, at the bottom of the sheet, applies to both, as in his MS. he quotes both Forster and Anderson for the species. I am thus particular in alluding to the British Museum specimens, because there is a discrepancy between the plant of Forster as described by him, and our own, according to his MS. description, published by M. A. Richard, l. c., where the tube of the corolla is described as being twice the length of the calycine segments, and the latter as subulate. In all our specimens, both from Lord Auckland's and Campbell's Islands, as also in those of Antarctic America, the tube of the corolla is a little longer than the calyx, sometimes as much as one-third, but it appears even more so before the expansion of the corolla ; and by subulate, that author might have alluded to the acuminated apex which the segments sometimes have. Though Forster's drawing does not exhibit the calyx, it coincides too closely with the preserved specimen, and both with our plant, to leave any doubt in my mind that we have here another instance of the similarity of the vegetation of the higher latitudes. Dr. Solander, indeed, considers the New Zealand plant as a different variety from the Southern American, and in his MS. description of the southern species, to which I have access through the kindness of Mr. Brown, he separates the former as ' *B. floribus carneis*

(Forster), ramis glabriusculis, frutex sesquipedalis.' In Forster's drawing, the mineral white, used to colour the flowers, has become discoloured ; and the pink, alluded to by Dr. Solander, almost obscured ; in our specimens they are of a pure milk-white, when fresh. The want of down on the branches arises from age.

"In Lord Auckland's group this species attains a much larger size than it does in America, there seldom exceeding four feet in height ; whilst Forster describes the Dusky Bay tree as twelve feet, and I have seen it as much as thirty on the margins of the woods close to the sea, where it may be readily distinguished by its pale green foliage and erect branches. I saw but one specimen in full flower, growing on an inaccessible rock, overlooking Rendezvous Harbour ; from a distance it looked powdered with white flowers."

279

572. EPACRIS NIVALIS. *Loddiges.* A half-hardy evergreen bush, from Australia. Flowers pure white. Belongs to Epacrids. (Fig. 280.)

This was introduced from New Holland by the late Henry Moreton Dyer, Esq., while vice-president of the Horticultural Society, who gave seeds of it, in 1829, to Mr. Loddiges, in whose Botanical Cabinet an excellent figure afterwards appeared. It forms an evergreen bush, which, when loaded like an Andromeda with hundreds of snow-white flowers, is exceedingly ornamental. Any greenhouse will afford it protection enough in winter, and in summer it will bear the open air of this climate. In Dorsetshire indeed it is found to be perfectly hardy ; Mr. Strangways having furnished us with the specimen from which our cut was taken, from his garden at Abbotsbury. In the open air it is very much handsomer than in a greenhouse, dwarf, compact, and crowded with little white bells, nestling among the black-green leaves. It is not unlikely to stand even a London winter if placed in a Northern exposure.

280

573. PAULOWNIA IMPERIALIS. *Siebold & Zuccarini.* A hardy deciduous tree, belonging to the Natural Order of Linariads. Native of Japan. Flowers violet and sweet-scented.

The Right Rev. the Lord Bishop of Exeter did me the favour to send me two panicles from his favoured grounds of Bishopstowe, near Torquay. "The blossoms," his Lordship writes, "are in terminal clusters ; and the odour (which will probably be lost when it reaches you) is of a very delicate violet-like character."—" But, after all, the effect to the eye is rather disappointing ; for the blossom precedes the leaves, which are not yet half out." The fragrance, so far from being lost on the journey, was rather increased, and the box retained the very agreeable odour some days after the flowers were removed. Unquestionably the absence of leaves, as the Bishop justly observes, is a great deficiency, especially in a plant whose size prevents the blossoms from being closely inspected upon the tree ; yet a cut panicle of these large pale violet-purple blossoms, as large as those of the Foxglove, with a young shoot of tender green leaves, is a very lovely object, to say nothing of the fragrance as a further recommendation. Unfortunately it is only the climates analogous to the south of Devonshire where its blossoms can be reasonably looked for. About London we find our strongest and healthiest plants with their terminal shoots (which alone produce flowers) nipped, and more or less killed, by the winter's cold, or, what is worse, the biting north-east winds of spring. The summer-growth of this tree is almost everywhere, in the middle and south of England at least, remarkable : stout limbs are thrown out in a short time, bearing ample foliage ; but these limbs are soft and succulent, the later shoots incapable of bearing a moderate frost. In France, even at Paris, the wood ripens better. Although forming a tree (in its native country, Japan, thirty to forty feet high), and bearing flowers like a Bignonia, and with a foliage and habit like Catalpa, the Paulownia belongs nevertheless to the Scrophularia family. Dr. Siebold considers it " un des plus magnifiques végétaux du Japon ;" and partly on this account and partly " parceque la feuille ornée de trois tiges de fleurs a servi d'armes au célèbre héros TAIKASMA, est encore aujourd'hui fort en honneur en Japon,"—" nous avons pris la liberté de nommer PAULOWNIA ce nouveau genre, pour rendre hommage au nom de Son Altesse Impériale et Royale la Princesse héréditaire des Pays Bas." In Japan the trunk of the tree attains an elevation of thirty to forty feet. Its growth in Dr. Siebold's garden has been six to ten feet in one year, and in three years a diameter of four to five inches. The flowers appear in April, and are grouped in large compound panicles, like those of the Horse-chestnut. It appears most abundantly in the southern countries of Japan, flourishing in the valleys and on the sides of hills exposed to the powerful action of the sun.— *Bot. Mag.*, t. 4666.

574. ACROPERA CORNUTA. *Klotzsch.* A brown-flowered epiphyte, from Guatemala. Introduced by Mr. Warczewicz. Flowered with Mr. Allardt, nurseryman, Berlin.

A. cornuta ; pseudobulbis cæspitosis ovatis, apice attenuatis bifoliatis ; foliis oblongis 3—5-costatis acuminatis, basi longe attenuatis, læte viridibus ; racemis basilaribus pendulis sesquipedalibus e viridi-purpureis 16—20-floris ; bracteis lanceolatis acuminatis ; perigonii foliolis exterioribus obovatis longius apiculatis fulvis, lateralibus tortis, supremo patentissimo recto, interioribus semilunato-lanceolatis brevibus ; labello brevi unguiculato saccato, apice longissime incurvo-cornuto ; gynostemio albido, intus ad basin purpureo-punctato ; germinibus leviter striatis pedicellisque purpurascentibus.

In its habit this plant much resembles *A. Loddigesii,* only the pseudobulbs, leaves, racemes and flowers, are larger. The racemes attain a length of one and a half or two feet, and the number of flowers on each raceme varies from sixteen to twenty-two. The colour of the flowers, which in *A. Loddigesii* are yellowish-brown inclining to green, is here pale yellow. The sepals are obovate, keeled at the back, and run out into a long soft horn. The bag-shaped labellum has the colour of yolk of egg, spotted with red on the inside, half an inch long, and on the outside towards the point is provided with an incurved horn four lines long.—*Allgem. Gartenzeit., June 12,* 1852.

575. COSCINIUM FENESTRATUM. *Colebrooke.* (*aliàs* Pereiria medica *Lindl. ; aliàs* Menispermum fenestratum *Gærtn. ; Wennewelle,* or *Wennewelle-gette,* of the Cinghalese.) A broad-leaved climbing shrub, with brownish-green flowers. Native of Ceylon. Belongs to Menispermads. Introduced at Kew.

We have received seeds of this plant at the Royal Gardens of Kew, from Mr. Thwaites, of the Botanic Garden in Ceylon. There has been of late a very extensive importation of what we here term "*false Calumba-root,*" instead of the true Calumba-root, *Jateorrhiza palmata,* Miers. Daniel Hanbury, Esq., of Plough Court, London, in a recent volume of the *Pharmaceutical Journal,* gave a history of this fraud on the public ; and immediately opened a correspondence with Mr. Thwaites on the subject of the plant in question. The *Coscinium* was scarcely known to botanists but by the brief description of the curious seed, and the still imperfect description of the plant by Mr. Colebrooke in the *Linnæan Transactions,* and Dr. Roxburgh in his *Flora Indica,* from specimens and information communicated to those Indian botanists from Ceylon by General Macdowall. A notion had prevailed, derived from the name of the Calumba or Columbo plant or root, that it was derived from Columbo in Ceylon, and a native of that island. At length it was ascertained that the true plant was a native of Mozambique, where it is known by the name of *Kalumb* or *Kalumba.* General Macdowall then sent out our present plant to his scientific correspondents in order to ascertain whether this, much celebrated in the Cinghalese Pharmacopœia, was not the *true* Calumba-root, and for that purpose consigned " a pretty large bit of the *root,*" sawed from the centre of a knot, to Dr. Roxburgh, that he might make experiments with it. Dr. Roxburgh, in a note, *Fl. Indica,* p. 811, at once sets the question at rest : " This is certainly not the *Calumba-root* of our Materia Medica." Nevertheless there have been large importations and ready purchasers for the Ceylon drug into England, the real properties or virtues of which (belonging though the plants do to the same Natural Family) are, to say the least, very problematical. It now only remains for us to give Mr. Thwaites's remarks and descriptions in his own words. " This species is very abundant near the sea-coast in Ceylon, and occurs also in the Central Province. The specimens from which the accompanying figure was taken were procured about twelve miles from Kandy. The Cinghalese value this plant very highly, using a decoction of the knotty parts of the stems (not the root) as a tonic and anthelmintic. The wood yields an inferior yellow dye." Some further remarks are given by Mr. Thwaites in a letter dated Peradenia, August 14, 1851 :—" The *Menispermum fenestratum* Roxb. is taken here, I am told by an intelligent native, *mixed with other things*, in a great many complaints, and applied externally in some cases, such as for weak eyes, &c. The mode of preparing it, is to chop up the wood at the knots of the stem very small, and to boil it (with other things, which was particularly impressed upon me) in seven measures of water, until they are evaporated down to one measure. It seems to be one of the universal medicines employed here in any and every complaint. It is quite impossible to get at any definite information from the natives as to what particular complaints certain plants are useful in. The priests, who are the doctors, appear to me to mystify the poor people by directing them to take certain leaves and roots, which it often gives them no little trouble to find ; and I think that the mind being employed in the matter, as well as the bodily exercise the patient often takes to procure the valued remedies, and a certain mixture of faith, have more to do with the cure than the drugs, some of which are evidently perfectly valueless, except to feed cattle." —*Bot. Mag.*, t. 4658.

576. GREVILLEA ACANTHIFOLIA. *A. Cunningham.* A half-hardy evergreen shrub from Australia. Flowers purple, in April and May. Belongs to Proteads. (Fig. 281.)

There is no doubt that some of the Proteads from New Holland are very nearly if not quite hardy. *G. sulphurea* and *rosmarinifolia* are open ground bushes at Exeter, and this, always regarded as a greenhouse plant, requires no pro-

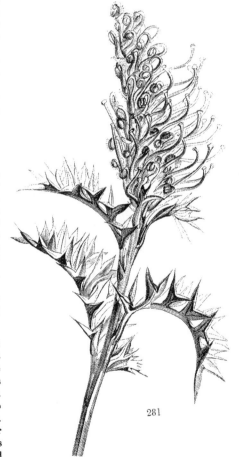

tection in Dorsetshire, where it flowers in the border among other shrubs, with the Hon. W. F. Strangways. The species is said to have been found by Allan Cunningham in peaty bogs on the Blue Mountains and banks of Cox's river during Oxley's expedition into the interior in 1817 ; and was shortly after raised at Kew. In the *Botanical Magazine* Dr. Graham has given the following description of the plant as he saw it in Mr. Cunningham's Nursery at Comely Bank, near Edinburgh :—" Shrub erect ; stem round, bark brown, branches scattered, angular, green. Leaves scattered, pinnatifid, rigid, smooth on both sides, revolute in their edges, dark green above, paler below ; pinnæ wedge-shaped at the base, trifid, segments tipped with a spine ; middle-rib of the leaf, pinnae and pinnulae prominent below. Racemes terminal upon short branches, opposite to the leaves, spreading. Flowers all turned upwards, refracted, sessile. Calyx lanato-sericeous on the outside, purple within and smooth, segments at length distinct, deciduous. Anthers dark red orange-coloured, after shedding the pollen yellow, bilocular, sessile. Germen stipitate, silky, lateral, gland on the anterior side of the base of the footstalk, lobular, semicircular, secreting abundance of honey. Style curved, quite smooth, and shining pink. Stigma flattened, set straight on the top of the style, green, or bursting from the calyx ; it carries on its centre a round and prominent mass of the dark-coloured pollen."

577. CEANOTHUS VERRUCOSUS. *Nuttall.* A hardy evergreen shrub from California, with pale bluish flowers. Belongs to Rhamnads. Introduced by the Horticultural Society.

The discovery of this pretty and, as it proves, hardy evergreen shrub is due to the venerable Mr. Nuttall, who found it at Santa Barbara, Upper California. Our plants are derived from the Horticultural Society, who appear to have received the seeds from Hartweg, while he was in California, under the name of " *C. integerrimus;* " but by that name he could not intend the plant so called of Hooker and Arnott, in the *Botany of Beechey's Voyage.* The plants have borne the open air in the Arboretum at Kew for two winters, and flower readily in April and May. Our specimens have been carefully compared with Mr. Nuttall's original ones, and they seem entirely to agree. The foliage in our plants is rather

281

larger and generally more orbicular, a change that may be due to cultivation ; and in both the leaves are very variable, even on the same specimens. Our flowers are very pale purplish-blue. They would appear " white " in the dried plant as described by Torrey and Gray. Our plant is nearly four feet high, much branched, with opposite and more or less spreading branches, which are terete, glabrous, studded at the nodes with two to four large, brown, ovate, acute, warty excrescences. Leaves opposite, and generally bearing a fascicle of young leaves in their axils, oval or cuneate, or orbicular-cuneate, or quite orbicular, almost sessile, very entire or more or less dentate, coriaceous, dark green, persistent, quite glabrous and glossy, and obscurely penninerved above, paler beneath, strongly penninerved and reticulated, the areolæ of the compact reticulations minutely villous. Corymb from the apex of small lateral branches : the rachis elongated, fleshy, indented as it were to receive the pedicels. Flowers pale purplish-blue. Calyx of five erecto-connivent ovate segments. Pedicels unguiculate ; the lamina cucullate. Stamens five : filaments subulate, nearly erect, opposite the petals. Ovary sunk in a fleshy disc, and surmounted by five lobes. Style thick. Stigmas three, capitate. Fruit in Mr. Nuttall's specimens as large as a small pea. —*Bot. Mag.*, t. 4660.

578. BEGONIA MONOPTERA. *Link and Otto.* A tuberous greenhouse perennial. Flowers pure white. Native of Mexico. Belongs to Begoniads. (Fig. 282.)

This very pretty species seems to be unknown in England. It was found in Mexico by Deppe, and by him the tubers were sent in 1826 to the Botanical Garden, Berlin, where it flowered. It is described as having a simple taper reddish stem, growing two feet high and more, and covered with extremely delicate vesicles. The leaves have a long stalk, which is flat towards the top ; its blade wedge-shaped, unequal-sided, three inches long and three inches broad, obliquely truncate, crenated in an irregular manner, bright green on the upper side, deep red on the under. The flowers grow in a terminal thyrse, with slightly downy flower-stalks. Among the flowers are some bulbs. The ovary has one lanceolate wing, three lines long. Both males and females have five petals, which are white, with the edge rolled back. The flowers appear in July and August. It is propagated by tubers, seeds, and the small tuber-like bodies among the flowers. The latter should be placed in dry sand as soon as the stems are dead.—*Link and Otto, Icones.*

579. DENDROBIUM FARMERI. *Paxton. Mag. Bot.* A beautiful hothouse epiphyte, with pink and yellow flowers. Native of the East Indies. Blossoms in May.

D. Farmeri (Dendrocoryne) ; caulibus elongatis clavatis articulatis profunde sulcatis basi pseudobulbosis apice foliosis, foliis 2—4 ovatis coriaceis striatis, racemis lateralibus multifloris pendulis, bracteis parvis ovatis concavis, sepalis (alboflavescentibus roseo-tinctis) late ovatis obtusis, petalis conformibus (ejusdemque coloris) majoribus, labello majore (pallide flavo disco luteo) rhomboideo obtusissimo unguiculato lato supra pubescente margine denticulato.—*Hooker.*

A most delicate and lovely Dendrobium, sent in 1847 by Dr. M'Clelland, from the Calcutta Botanic Garden, to W. G. Farmer, Esq., after whom it was named. Mr. Paxton observes, that "in habit and appearance the plant very much resembles *Dendrobium densiflorum,* but the stems are more angular, and the flower-scape is less densely laden with bloom ; the flowers, too, are altogether different." The flowers, however, are more different in colour than they are in shape ; and if true to its other characters, there is no difficulty in distinguishing this species. In the stove of the Royal Gardens of Kew it flowers in May. Our plant has elongated club-shaped stems, jointed and deeply sulcated, growing in clusters ; at the base they swell out into a kind of pseudobulb, scarcely so large as a hazel-nut. The young stems bear from two to four spreading, ovate, coriaceous or fleshy leaves at the top, acute, striated ; the old stems throw out pendulous racemes from near the summit, which exceed the stems in length. Flowers numerous, but rather lax. Bracteas small, ovate, concave. Sepals very patent, broad, ovate, obtuse, pale straw-colour, delicately tinged with rose. Petals of the same colour and form, but larger, spreading. Lip moderately large, pale straw-colour, the whole disc

282

orange-yellow, broadly rhomboid, very obtuse, downy above, the base contracted into a claw, and above the claw the margin is on both sides folded and sinuated : the base above bears an oblong flattened tubercle. Column very short, terminated by the obtusely conical anther-case : the lower part of the column is extended downwards, so as to form an obtuse spur to the labellum.—*Bot. Mag.*, t. 4659.

This beautiful species is very near *D. chrysotoxum*, from which it differs in its lip not being so much fringed, nor so large, and in the sepals being suffused with pink.

283

580. POSOQUERIA REVOLUTA. *Nees v. Esenbeck.* A hothouse shrub. Flowers very long, white, sweet-scented. Belongs to the Order of Cinchonads. Introduced by Messrs. Veitch and Co. (Fig. 283.)

This handsome shrub was produced by Messrs. Veitch, in April last, at one of the meetings of the Horticultural Society. The leaves are evergreen, ovate-oblong, rather acuminate, with a stalk about half an inch long, and the edge slightly rolled back. The flowers are five or six together, on smooth stalks about a quarter of an inch long, and furnished at the base with two extremely minute sharp scales ; they gradually taper into the ovary, which is surmounted by five sharp triangular teeth. The tube of the corolla is four inches or more long, very slender, and suddenly expands into a five-lobed limb, the divisions of which are linear-obtuse, and not more than three-quarters of an inch long. It seems to be the same as No. 767 of the Vienna distribution of Pohl's plants.

581. CORYANTHES SPECIOSA. *Hooker.* (*aliàs* Gongora speciosa *Hooker.*) A stove epiphyte from Brazil. Flowers very large, pendulous, dull pale yellow. Introduced about 1825. (Fig. 284.)

This, the first of the Coryanths that was discovered, was originally supposed to have erect flowers, and is so represented in the Botanical Magazine ; but in fact they are pendulous, and necessarily so, as in all the others. From the branches of trees on Victoria Hill, above Bahia, hang down little vegetable buckets, into which a pair of stumps or fingers constantly distil a sweetish colourless fluid, which, drop by drop, gradually fills the bucket. The fingers are processes springing from the base of the column of this Orchid, the bucket is a great helmet-shaped lip, sustained by a stiff arm (unguis) which keeps it perfectly steady, so that the honey may not be spilt. The column itself turns back as if to keep its head out of the way of the drops, while the broad membranous lateral sepals, resembling bats' wings, turn quite back, as if to unveil the singular phenomena which the blossoms present. The plant is very rare in collections ; it is easily known by the absence of spots from its flowers, and by the narrow neck which connects the helmet with the globular hood (intervening between it and the stalk) not being plaited. The glandular finger-like processes are moreover remarkably short. The smell of the flowers is to us rather unpleasant, some think it agreeable. Other Coryanths in cultivation are *macrantha, Albertiniæ, maculata,* and *Fieldingii.* Of the present there are two varieties, one with pale yellow, the other with almost white flowers.

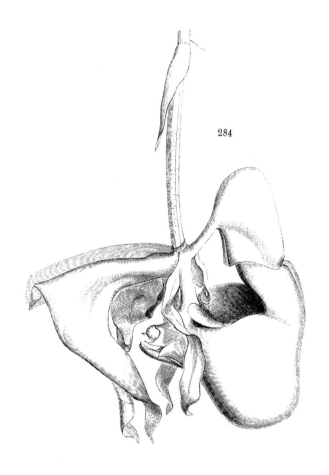

284

PLATE 94.

L.Constans del.& zinc.

Printed by C.F.Cheffins, London.

[PLATE 94.]

THE WOOLLY CLEMATIS.

(CLEMATIS LANUGINOSA.)

———◆———

A very fine large-flowered Hardy Climber, from CHINA, *belonging to the Order of* CROWFOOTS.

═══════════

Specific Character.

THE WOOLLY CLEMATIS. Leaves simple and ternate ; leaflets coriaceous, cordate, acuminate, shaggy on the underside as are the footstalks. Buds, peduncles, and young leaves buried in wool. Sepals six, ovate, acuminate, spreading flat.

CLEMATIS *LANUGINOSA* (VITICELLÆ) ; foliis simplicibus ternatisque, foliolis coriaceis cordatis acuminatis subtùs petiolisque villosis, alabastris pedunculis foliisque junioribus lanatis, sepalis 6 ovatis acuminatis patentissimis.

═══════════

THIS magnificent plant flowered last spring in the nursery of Messrs. Standish and Noble of Bagshot, who received it from Mr. Fortune. We have a wild specimen from that enterprising traveller, marked "Hills of Chekiang, July, 1850," and he has also favoured us with the following memorandum concerning it :—

"This pretty species was discovered at a place called Tein-tung, near the city of Ningpo. It is there wild on the hill sides, and generally plants itself in light stony soil near the roots of dwarf shrubs whose stems furnish it with support as it grows. Before the flowering season arrives it has reached the top of the brushwood, and its fine star-shaped azure blossoms are then seen from a considerable distance rearing themselves proudly above the shrubs to which it had clung for support during its growth. In this state it is most attractive, and well repays any one who is bold enough to scramble through the brushwood to get a nearer view.

"The flowers of this species are much larger and more hairy than those of the Japanese *C. azurea grandiflora,* to which it bears some resemblance. It is no doubt equally hardy, and perhaps more so. As a neat pot-climber for the greenhouse it will be much prized. The situations and soil in which it is found wild will point out the true mode of managing it in our gardens."

It is no doubt very near *C. azurea,* from which it differs in the leaves being coriaceous not thin, shaggy beneath with white hairs not finely silky, and cordate not ovate; in the flower-buds, young leaves, and peduncles being buried in wool, not subpubescent, and in the great size of the flowers, whose divisions are broader and more acute.

There seems no reason to doubt that it is quite hardy; and if so it will prove one of the finest, if not the finest, climber our gardens yet possess.

PLATE 95.

L.Constans del.& zinc.

Printed by C.F.Cheffins, London.

[PLATE 95.]

THE BEAUTEOUS VERONICA.

(VERONICA FORMOSA.)

———◆———

A handsome Evergreen Half-hardy Shrub, from VAN DIEMEN'S LAND, *belonging to* LINARIADS.

══════════════════

Specific Character.

THE BEAUTEOUS VERONICA. Shrubby. Branches hairy in two lines. Leaves on very short stalks, oblong-lanceolate, acute, entire, one-nerved, narrowed to the base, smooth. Racemes few-flowered, loosely corymbose at the ends of the young branches. Segments of the calyx narrowly lanceolate, acute. Capsule twice as long as the calyx.

VERONICA *FORMOSA ;* fruticosa, ramis bifariàm pilosulis, foliis brevissimè petiolatis oblongo-lanceolatis acutis integerrimis uninerviis basi angustatis glabris, racemis in apicibus ramulorum paucifloris laxè sub-corymbosis, calycis segmentis angustè lanceolatis acutis, capsulâ calyce duplò longiore.—*Bentham.*

Veronica formosa : *R. Brown, Prodr.,* 434 ; *Bentham, in De Candolle's Prodromus,* 10, 462 ; *aliàs* V. diosmæfolia : *Knowles and Westcott, Fl. Cab.,* 3, 65, t. 106.

══════════════════

A NATIVE of Van Diemen's Land, where it appears to be plentiful. Mr. Gunn, from whom we have wild specimens, says "it is common on the South Esk at Launceston, and I have gathered it at an altitude of 3500 feet above the sea on Mount Wellington and the Western Mountains. Are there not two species under my number 527 ?"—a question which he is more competent to pronounce upon than we are.

What we have in cultivation is a compact, dark green, evergreen bush, with small box-like leaves arranged in a distinctly decussate or four-rowed manner, and always having a great tendency to preserve the horizontal line, or even to curve below it. The flowers are a clear bright blue, appear in little corymbs at the ends of the branches, and are much like those of *V. maritima* on a large scale.   Their size varies much, according to the health of the plant and the climate in which they are produced.   Those now represented, from a plant which lives out of doors without protection, at Abbotsbury, in Dorsetshire, with the Hon. W. F. Strangways, are by no means so large as we find upon some of the wild Van Diemen's Land specimens.

The Mount Wellington plant, alluded to by Mr. Gunn in the above memorandum, we also possess from Mr. George Everett.   It has narrower leaves, more strikingly recurved than in Launceston specimens, and smaller flowers; but these are differences that may very well be caused by an alpine climate.

Mr. Strangways' plants flower with him in June: when the bushes become objects of great beauty.   It is not, however, to be expected that the species will be hardy in the midland counties. There it would probably live in a glass wall, a proper place to try it in: or even beneath a north wall out of the way of direct sunlight; but this is to be determined experimentally.   Near London it is regarded as a greenhouse plant.

The name *V. diosmæfolia* applied to this by Messrs. Knowles and Westcott, and still retained here and there in gardens, belongs to a totally different shrub, with small white flowers, from New Zealand.

PLATE 96.

L.Constans del.& zinc.

Printed by C.F.Cheffins, London.

[PLATE 96.]

THE PURPLE-STAINED LÆLIA.

(LÆLIA PURPURATA).

———◆———

A magnificent Stove Epiphyte, from St. Catharine's *in Brazil, belonging to the Order of* Orchids.

═══════════════════

Specific Character.

THE PURPLE-STAINED LÆLIA. Pseudobulbs oblong. Leaves narrowly oblong, emarginate. Peduncles two-flowered, proceeding from a spathe. Sepals linear-lanceolate ; petals oblong-lanceolate, obtuse. Lip very large, rolled round the column, rounded, the lateral lobes very obscure and hardly distinguishable from the middle one.

LÆLIA *PURPURATA ;* pseudobulbis oblongis, foliis angustè oblongis emarginatis, pedunculis bifloris è spathâ erumpentibus, sepalis lineari-lanceolatis, petalis oblongo-lanceolatis obtusis, labello maximo circa columnam convoluto rotundato lobis lateralibus obsoletis ab intermedio parum diversis.

═══════════════════

One of the most striking novelties which has for a long time been seen was produced by Messrs. Backhouse of York, at one of the garden meetings of the Horticultural Society, under the name of a new Cattleya from the island of St. Catharine's in Brazil. It had in fact much the appearance of *Cattleya crispa,* or of a white *C. labiata,* but the experienced eye of one of our most acute Orchidophilists suggested to him at the first glance that it was probably a Lælia related to *L. Perrinii.* And such it proved to be when the pollen-masses were examined ; they are eight, not four.

The pseudobulbs are oblong, and produce at their end a narrow oblong blunt leaf, as broad at one end as the other, about eight inches long, and deeply notched at the point. In the axil of the leaf comes a compressed pale green spathe fully three inches long, and much like that of *Cattleya labiata.* The peduncle which appears from within this is stout, deep green, and two-flowered. The flowers are rather more than six inches from the tips of the petals. Sepals and petals pure white ; the former linear-lanceolate, rolled back at the edge towards the base and thus appearing unguiculate; the latter three times as broad, ovate-oblong, obtuse, wavy. The lip is three inches long, rolled round the column, with a much-rounded point from which the rounded lateral lobes are hardly distinguishable; it is yellow in the middle towards the base and streaked with crimson, but the limb is of the deepest and richest purple, diminishing in intensity towards the edge.

It is evidently very near the *Lælia grandis*, another Brazilian species, introduced into this work at No. 21 of the Gleanings; but that species is represented to have a leaf broader at the base than the point, and nankin-coloured flowers, with a white lip washed with rose at the base; the sepals and petals are also narrower, more wavy, sharper, the latter serrulate, and the lateral lobes of the lip very distinct and ovate.

The vignette represents the plant as it was exhibited at Chiswick.

GLEANINGS AND ORIGINAL MEMORANDA.

582. DENDROBIUM BARBATULUM. *Lindley.* A handsome epiphyte from Bombay. Flowers white. Introduced by Jas. Bateman, Esq. (Fig. 285.)

This species is by no means rare in collections, under the erroneous name of *D. Heyneanum,* which it has acquired, Heaven knows how, the real plant of that name being totally different. *D. barbatulum* was originally taken up, from indifferent materials, out of Heyne's Herbarium, distributed by Dr. Wallich. It was afterwards imported from Bombay, whence I received it in 1844 from Mr. Bateman, under the name of *D. Heyneanum.* It often appears at exhibitions, where it is known by its erect spikes of pure white muslin-like flowers, in which not a tinge of any other colour is visible. The

285

sepals and petals are much alike, lanceolate and acute, but the petals are the broader of the two. The lip is three-lobed, very slightly downy, with two short lateral obtuse lobes, and a linear callosity reaching as far upwards as the sinus of those lobes. The middle lip obovate and obtuse. A transposition of labels, memoranda, and sketches led me into the great error of confounding this with the widely different *D. chlorops.* (See *Bot. Reg.,* 1844.)

583. LANSBERGIA CARACASANA. *De Vriese.* A stove tuberous-rooted plant. Flowers golden-yellow spotted with black. Native of the Caraccas. Belongs to Irids. Introduced to the Botanic Garden of the University of Leyden.

This curious plant, although it flowered six years ago at Leyden, does not seem to have yet reached England. It was obtained in the Caraccas by M. Reinhardt van Lansberg, a Dutch gentleman, who sent many fine things in 1845 to the Botanic Garden of that university. Professor De Vriese describes it as having the habit of *Marica, Phalocallis, Cypella, Moræa,* &c. The root tuberous. Stems simple, compressed, zigzag, tumid at the joints, half a yard long. Radical leaves equitant, distichous; stem-leaves sheathing, compressed, from three to five times shorter than the others. Spathes terminal, compressed, leafy with pellucid membranous edges. Sepals largest, expanded from a narrow base, then contracted, and then widened again; from the base to the middle contraction spotted with brown or black upon a golden-yellow ground. Petals somewhat panduriform, with an ovate chesnut-brown spot above the middle contraction almost bordered by a yellow ground, with two minute brown spots near the edge. It is said to flower all the year round in the stove, one flower only appearing at a time, and very fugacious. The learned author of the genus observes that *Phalocallis* has a goblet-shaped flower with spreading sepals; *Lansbergia,* on the contrary, has all the sepals closed and converging, besides which its leaves are neither plaited nor ribbed. In *Phalocallis* the cells of the anther are attached by the upper part only to the lobes of the style, but in *Lansbergia* they adhere by their whole length. In the former the stigmata are transversely two-lobed, in the latter they are minutely crested, and by no means petaloid. *Cypella* differs in its "stigmas being distinctly lobed, acute, stretched forward, horny, fringed at the upper side with acute horny crests." The scientific characters of this novelty are thus given in the *Epimetron ad indicem seminum anni* 1846, *de plantis novis in hort. bot. Ac. Lugd. Bat. cultis.*

LANSBERGIA. *De Vriese.* Perianthium superum, hexaphyllum, petalis dimorphis, tribus externis majoribus concavis, internis angustis, apice involutis. Filamenta basi vix connata, filiformia, loculis antherarum latere dehiscentibus, styli lobis dorso per totam longitudinem adglutinatis. Stylus trigonus, apice trilobus, lobis clavæformibus, stigmatibus (loborum apicibus) brevissimis, vix cristatis. Capsula triquetra-oblonga, operculo vix conspicuo. Semina globoso-angulata, scrobiculata.

L. CARACASANA. *De Vriese.* Foliis equitantibus, elongatis, ensiformibus, ancipitibus, caulem sesquipedalem vix superantibus, spatha multivalvi, compressa, 2—4 flora, pedunculo trigono; perigonio læte aureo-flavescente, fugacissimo, laciniis exterioribus obovatis, acutis, brevissime mucronulatis, infra medium late nigro-maculatis, interioribus oblongis medio contractis et macula oblonga, nigra tinctis, apice dilatato, rotundatoque subtilissime mucronulatis; filamentis basi, badia primum conjunctis, demum solutis, flexuosis; antheris oblongis.

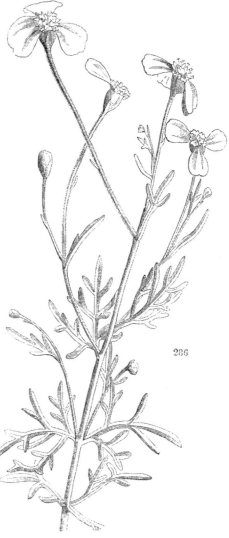

286

584. ACHYROPAPPUS SCHKUHRIOIDES. *Link & Otto.* A yellow-flowered inconspicuous annual. Native of Mexico. Belongs to the Composite Order. Introduced by the Royal Botanic Garden, Berlin. (Fig. 286.)

The name of this plant appearing in some seedsmen's lists of annuals, it is as well that it should be better known than it is. From Mexico it was sent to Berlin by Cervantes, the director of the Botanical Garden in that city. It is a small spreading annual resembling a Tagetes. Leaves bipinnatifid with linear segments, clothed with a few short scattered hairs. Flower-heads on long naked stalks. Involucre composed of imbricated carinate scales. Receptacle naked. Florets of the ray from one to three, wedge-shaped. It is not worth anybody's growing. See *Link and Otto's Icones, p.* 59, *t.* 30, from which our cut is borrowed.

585. BRASSIA KEILIANA. *Reichenbach fil.* A yellow-flowered epiphyte, with large keeled bracts. Belongs to Orchids. Flowers produced at Leipsic in the collection of M. Keil.

B. spicâ pauciflorâ, bracteis navicularibus, acutis, argutè carinatis ovaria superantibus, perigonii phyllis lineari-lanceolatis acuminato-aristatis, lateralibus internis brevioribus; labello a basi brevissime lateque cuneato oblongo, margine hinc microscopicè denticulato, undulato, apice acuminato, phyllis lateralibus internis breviore, lamellis baseos erectis utrinque obtusangulis, præsertim basin imam versus papillis velutinis, anticè in auriculas extrorsas appositas excurrentibus, gynostemio genuino, androclinii margine postico protenso, marginibus membranaceis foveæ stigmaticæ productis.—*Reichenbach fil. in litt.*

Sepals yellow, then beautifully brownish orange; cinnabarine when dried (as is *Miltonia flavescens* Lindl.); lip whitish. Named after Hofrath Keil of Leipsic, in whose garden it flowered. This gentleman has a very good collection of Orchids, consisting of large vigorous specimens, well managed by his skilful gardener Tube, who has also made some good experiments on exposing Mexican Orchids to the open air in summer.—*Rchb.*

We are indebted for our knowledge of this plant to Mr. H. G. Reichenbach the younger, of Leipsic. It is evidently allied to *B. glumacea,* but its flowers are much larger, and the lip of a different form.

586. MECONOPSIS WALLICHII. *Hooker.* A beautiful blue-flowered herbaceous half-hardy plant. Native of the Sikkim Himalaya. Belongs to Poppyworts. Introduced at Kew.

A very handsome species of Meconopsis, detected in Sikkim-Himalaya by Dr. Hooker, who sent seeds to the Royal

Gardens, which produced flowering plants in June, 1852. It is assuredly no described species, though agreeing in some respects with *M.? Nepalensis* De Cand. (*Papaver paniculatum* of Don), which has yellow flowers, and a " globose capsule, as large as a garden cherry." It quite accords with an unnamed " Meconopsis, n. 8123, β," of *Wallich's Catalogue*, from " Kamaon ?" Dr. Hooker has another and apparently distinct species in his Herbarium, with much longer yellow flowers, and a much more compound raceme, or panicle. The plant, with us, grown in pots in a frame, attains a height of two and a half to three feet : the whole herb is pale subglaucous green, everywhere hispid, with long spreading ferruginose setæ. Radical leaves large, petiolate, lyrato-pinnate, or pinnate below and pinnatifid above, the pinnæ and lobes ovato-oblong, sinuated. Stem-leaves sessile, oblong, pinnatifid. Flowers large, drooping, arranged in an elongated leafy raceme, compound below. Peduncles and pedicels rather short, curved downwards, erect in fruit. Calyx of two oblong, very concave, deciduous sepals. Corolla of four subrotundo-obcordate, spreading, pale-blue petals, having sometimes a slight tinge of green. Stamens very numerous. Anthers orange-yellow, crowded so as to form a large ring around the style. Ovary elliptical-oblong, clothed with a dense mass of erect, appressed, rufous, somewhat plumose setæ, one-celled, with six or seven parietal receptacles. Style cylindrical, as long as the ovary. Stigma capitate, of six or seven dark green erect lobes.—*Bot. Mag.*, t. 4668.

587. SCHLIMMIA JASMINODORA. *Planchon & Linden.* An orchidaceous epiphyte from Central America. Flowers white and very fragrant. Introduced by Mr. Linden. (Fig. 287.)

In Mr. Linden's interesting catalogue of 1852, we find the following notice :—

"SCHLIMMIA JASMINODORA. *Planch. et Lind.* Genre nouveau des plus curieux, à sépales inférieurs soudés ensemble et formant un sac ressemblant à ceux des *Cypripedium*. L'espèce en question porte une hampe inclinée de huit à dix pouces, garnie de dix à quinze fleurs d'un blanc pur, à odeur de jasmin fortement prononcée. Elle croit épiphyte et terrestre dans les forêts des versants tempérés de la province d'Ocana, où elle a été découverte par M. Schlim.—30 à 50 francs."

Specimens and a drawing, with which we have been favoured by Mr. Linden, leave no doubt about this being an entirely new genus of the subdivision *Vandeæ*. The plant appears to have a long taper slender pseudobulb, bearing a single long-stalked thin oval leaf. The scape, which is radical, is altogether a foot high, with about six distant loose oblong scales, and three secund flowers, each of which is about an inch long, pure white, with the lower sepals very large and grown into a deep bag, beyond which project a pair of linear reflexed petals. Although not a showy plant, its very fragrant flowers render it one very desirable in a hothouse. The genus may be characterised thus :—

SCHLIMMIA ; sepala carnosa, inæqualia ; dorsale lineare rectum liberum, lateralia maxima in saccum altum omninò connata. Petala sepalo dorsali æqualia, reflexa. Labellum minutum, ungue carnoso cum pede columnæ articulato tuberculato, limbo simplici membranaceo duplò breviore. Columna semiteres, apice utrinque auriculata, in pedem cum sepalis lateralibus connatum producta ; rostello setaceo deflexo. Pollinia 2, cereacea, caudiculâ elongatâ cuneatâ glandulâ minutâ lunatâ.

The lip is a fleshy body, shorter than the column, and articulated with it, with three knobs near the foot of the column, another in the middle of its length, and a fourth which is concave at its extremity, which is prolonged into a thin trowel-shaped limb.

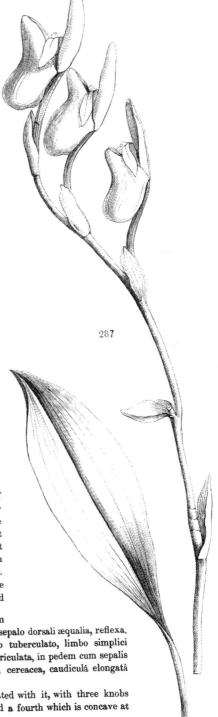

287

588. ERIA FLORIBUNDA. *Lindley; var.* leucostachya. An epiphyte from Borneo, with long close spikes of white flowers. Introduced by Messrs. Low and Co. of Clapton. (Fig. 288.)

This very pretty plant was exhibited in November, 1847, to the Horticultural Society by Mr. Low, whose son found it on the banks of the Sarawak River, growing in large masses on trees, the branches of which were fully exposed to the sun. The flower-spikes were said to be frequently from eight to ten inches in length. It was then named provisionally *E. leucostachya*: a suspicion being expressed that it might prove, when better known, to be a mere variety of *E. floribunda.* We have now received a finely-grown specimen from Mr. John White, gardener to A. Kenrick, Esq., of West Bromwich near Birmingham, and we are able to confirm that suspicion, the structure of the two plants being the same. The flowers are, however, much more closely packed, and pure white, without a tinge of the purple with which the thin spikes of *E. floribunda* are suffused.

589. MALCOLMIA LITTOREA. *R. Brown.* (*aliàs* Hesperis littorea *Lamarck; aliàs* Cheiranthus littoreus *Linnæus.*) A hardy annual, with large purple flowers. Belongs to the Cruciferous Order. Native of the South of Europe. Blossoms in the Autumn.

Of this really beautiful hardy plant, cultivated in our gardens so early as 1683, no good figure has hitherto been given. It is a littoral plant of South Europe. Its northern limit seems to be Nantes, and thence it extends itself along the coasts of Spain and Portugal, and the western shores of the Mediterranean. Desfontaines detected it in Barbary, and Broussonet in Morocco. In our country it is best treated as an annual. In warmer climes it is at least biennial, the lower part of the stems becomes quite woody, and then the branches are more strictly erect, and more numerous from one point than our figure represents them. Mainly on this account, as it would appear, Boissier makes two varieties, his var. *Broussonetii*, and var. *alyssoides*. Seeds were sent to us by Mr. Wellwitzsch from Portugal, and the plants bear their lovely flowers during the summer and autumn. Our annual plants (and they would hardly

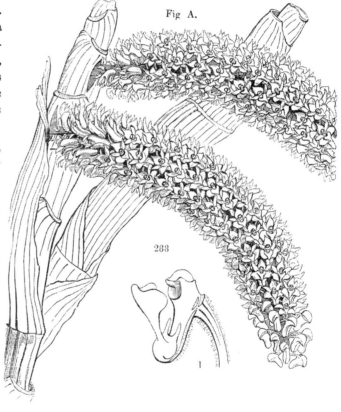

Fig A.

288

survive a winter in our climate) have erect, but flexuose, branching stems, scarcely a foot high, terete, hoary, as is the whole plant, petals and stamens excepted, with short stellated hairs. Leaves lanceolate or linear-lanceolate or more frequently subspathulate, tapering a good deal at the base, but sessile, sometimes sinuate-dentate, more usually quite entire. Flowers large for the size of the plant, in lax, terminal, many-flowered racemes. Pedicels at first very short, at length about equal in length to the calyx. Calyx narrow, oblong. Sepals linear, obtuse, quite erect, two of them a little gibbous at the base. Petals obcordate, clawed, delicate, bright pink-purple (not *albido-flavi*, as De Candolle describes them), the lamina spreading horizontally (not veiny, like *Malcolmia maritima*, Bot. Mag. t. 166). Stamens six : the four longer nearly equalling the pistil ; two shorter rather longer than the germen. Germen cylindrical, downy. Style short. Stigmas two, long, linear, glandular within, and at the margin and apex, and united for the whole length of their faces into one, more or less bifid at the point. Siliqua two or two and a half inches long, slender, terete (not torulose), flexuose, erecto-patent, terminated by the style and now sharp withered stigma.—*Bot. Mag.*, t. 4672.

590. MEDINILLA SIEBOLDIANA. *Planchon.* A fine white-flowered hothouse shrub. Native of

Java. Blossoms in September. Belongs to Melastomads. Introduced by Mr. Van Houtte. (Fig. 289.)

A brief notice of this will be found in our Vol. 1, p. 124, no. 176. The introduction of the plant to our gardens enables us to give a figure of it, from a specimen exhibited at the last July meeting of the Horticultural Society by Mr. Cole, gardener to J. Colyer, Esq., of Dartford, one of the most zealous and suc-

289

cessful competitors at the metropolitan summer exhibitions. It is a stiff erect shrub, with short coarse fibres in the place of stipules. The leaves are oblong, a little tapering to either end, thick, entire, triple-ribbed, pale on the under side. Panicles naked, erect, pyramidal. Flowers tetramerous. Petals white. Stamens purple. A very handsome stove plant, bearing carriage well, and therefore suited to the purpose of exhibition.

591. ANIA LATIFOLIA. *Lindley.* (*aliàs* Calanthe viridi-fusca *Hooker.*) A stove terrestrial plant. Flowers greenish brown. Native of Assam and Sylhet. Belongs to the Order of Orchids.

This plant, and another of the same genus, was distributed by Dr. Wallich under the name of *Ania latifolia* (*Wall. Cat.*, No. 3741), which was afterwards published in the *Genera and Species of Orchidaceous Plants*, p. 130. A third species was afterwards figured in the *Botanical Register*, 1844, t. 8, from Ceylon, as *Ania bicornis.* They constitute a little group nearly allied to Bletia and Phaius, from which the spur on the one hand, and the 3-lobed lip accompanied by a 6 or 8-celled anther, distinguish them on the other. They are all terrestrial tuberous plants, with solitary somewhat ribbed leaves, and long spikes of dull-coloured flowers. By some oversight that now mentioned has been referred to Calanthe in the *Botanical Magazine*, t. 4669, where we have the following account of it :—

" A native of Assam, whence it was sent to the Royal Gardens of Kew by Mr. Simon. It flowered with us in April, 1852 ; and is remarkable among known species of Calanthe for the erect or nearly closed sepals and petals, the peculiar form of the lip, and the colour of the flowers. We presume it to be terrestrial. The habit approaches that of *Calanthe Masuca* more than any other species. The pseudobulb is broad-ovate, spreading out most on one side, dark green, firm, at the base furrowed and lobed, the upper part more or less covered with the remains of the long sheathing scales of the preceding year's leaf. Leaf solitary, arising from an infant inconspicuous pseudobulb, a foot or more long, lanceolate, membranaceous, plicato-striate, much and gradually acuminated at the point, the base tapering into a very long petiole, which is sheathed by three or four long cylindrical scales. Scape (including the long lax spike) a foot and a half long, terete, glabrous, erect, arising from the base of a pseudobulb bearing brown, striated, sheathing, membranous bracteas, especially at the base. Spike many-flowered, bracteated ; bracteas subulate, green, one under each ovary, and shorter than it. Ovary slender, clavate. Flowers greenish brown, moderately large. Petals and sepals lanceolate, nearly uniform, and, as well as the labellum, erect, so as almost to close over the column of fructification, quite concealing it. Lip broad, oblong or oblong-spathulate, applied to the column, but scarcely connate with it, which is embraced and almost included in its involute sides ; three-lobed, lateral lobes ovate, erect, middle or terminal one a little reflexed, cordato-subrotund, mucronate ; the colour of the lip is yellowish green, spotted or dotted in lines with purple within ; and, running nearly the whole length of the disc, are three lamellæ, a little fimbriated at their termination. Spur short, blunt, compressed, incurved, yellow, didymous at the apex. Column long for the genus, semiterete, furrowed in front, yellowish, blotched with rose-colour. Anther-case sunk in the apex of the column. Pollen-masses eight, as in the genus."

592. ONCIDIUM QUADRICORNE. *Klotzsch.* A species of unknown origin, with panicles of small brownish yellow flowers. Observed in blossom in the nursery of M. Allardt of Berlin.

Oncidium (Euoncidium §§. Labellum panduratum, medio constrictum) quadricorne *Kl.* Pseudobulbis lenticularibus parvis monophyllis ; foliis carnosis lineari-oblongis recurvis acutis subsessilibus, dorso carinatis, basi attenuatis conduplicatis ; paniculis basilaribus erectis filiformibus ; perigonii foliolis patentissimis oblongo-obovatis sordide flavido-fuscescentibus, labello elongato pandurato stricto albido apice bifido, lobis lateralibus obsoletis, crista basilari erecta alba quadridentata ; columnæ nanæ candidæ alis erectis ovatis.

The foregoing character is given in the *Allgemeine Gartenzeitung*, Aug. 7, 1852. The fleshy leaves are said to be linear-oblong, recurved, acute, and placed singly on lenticular pseudobulbs. The flowers are in slender panicles, dirty yellow, with a long whitish lip, and a white four-toothed crest. It seems to be very near *O. Harrisonianum.*

593. ALLARDTIA CYANEA. *Dietrich.* A blue-flowered stove herbaceous plant, native of Guatemala. Belongs to the Bromeliaceous Order. Introduced by M. Allardt of Berlin.

Dr. Dietrich has named this, which he conceives to be a new genus of plants, after M. Allardt of Berlin, who is said to have the finest trade collection of Orchids in Prussia. It is described as being a simple-stemmed Bromeliaceous plant with a branching panicle of green and blue flowers, growing from the centre of a rosette of strap-shaped entire leaves. The whole plant when in flower is said to be two and a half feet high. Each flower lasts for a day. The following are the characters assigned by Dr. Dietrich to the new genus and species.

Perigonium sexpartitum, laciniæ exteriores calycinæ, cum disco hypogyno turbinato connatæ, interiores petaloideæ, in tubulum convolutæ, liberæ, basi nudæ, apice patentes. Stamina sex, disco inserta ; filamenta filiformia, libera ; antheræ incumbentes, basi sagittato-emarginatæ. Germen disco turbinato insertum, liberum, pyramidatum, triloculare ; stylus filiformis ; stigma trifidum, lobis filiformibus spiraliter contortis. Fructus ?—

Allardtia cyanea. Herba americana, caulescens, simplex. Folia ligulato-lanceolata, integerrima, nuda, basi dilatata. Flores paniculati ; panicula ramosissima, ramis racemosis, spatha suffultis, ramulis spicatis, bracteatis.—*Allgem. Gartenzeit.*, 31 *July*, 1852.

594. GRINDELIA SPECIOSA. *Hb. Bentham.* A hardy undershrub, with large yellow flower-heads. Native of Patagonia. Belongs to Composites. Introduced by Henry Wooler, Esq. (Fig. 290.)

G. speciosa ; suffruticosa, viscosa, glabra. foliis oblongis basi angustatis inæqualiter inciso-dentatis, capitulis solitariis pedunculatis, involucro hemisphærico subsquarroso glutine copiosissimo viscosissimo obducto, receptaculo plano, pappi setis rigidis circiter 10 aliis corollæ longitudine aliis multò brevioribus.

This novelty was introduced by Henry Wooler, Esq., of Upper Tulse Hill, from whom we received it in the beginning of August last. He obtained the seed from his son, then at the Falklands, who had gathered it at a place called New Bay, on the coast of Patagonia, from a plant growing in the sand just above high-water mark. A specimen marked *G. speciosa,* collected by Captain Middleton in Patagonia, exists in Mr. Bentham's Herbarium. With Mr. Wooler it forms a bushy plant, two feet high, with from thirty to forty flower-heads open upon it at the same time. These heads are covered to a considerable thickness with a transparent glutinous varnish, by which this species is at once known. The obovate acutely and irregularly dentate leaves are also very viscid. It seems most nearly related to the Brazilian *G. buphthalmoides.*

595. EPIDENDRUM GUATEMALENSE. *Klotzsch.* A handsome species from Guatemala. Flowers yellowish green, dotted with purple, and with a white lip. Introduced by M. Allardt of Berlin. Blossoms in July.

Epidendrum (Encyclium) Guatemalense *Kl.* Caule adscendente subramoso radicante pseudobulboso ; pseudobulbis ex ovato-oblongis teretibus, apice attenuatis diphyllis ; foliis lineari-lanceolatis longis attenuatis brevi acutis coriaceis subtortuosis subtus carinatis racemum subsimplicem terminalem subduplo brevioribus : perigonii foliolis patentibus viridibus, extus intusque striis punctiformibus minutissimis fusco-violaceis ornatis, exterioribus oblongis utrinque attenuatis, interioribus spathulatis brevissime acutis ; labelli omnino liberi trilobi candidi lobis lateralibus obovatis columnam amplectentibus, intermedia orbiculari deflexa brevissime acuta, lineis violaceis angustis parallelis notata, basi angusta naviculari subcallosa ; columna trigona auriculata, auriculis obtusis inflexis vitellinis ; germinibus teretibus albido punctato-scabris.

According to Dr. Klotzsch who has described this in the *Allgemeine Gartenzeitung,* Aug. 7, 1852, this has taper pseudobulbs, each having two linear-lanceolate leaves, from eleven to eighteen inches long, and from half an inch to one inch broad. The panicle is two feet long, as thick as a crowquill, and carries from twenty to twenty-four flowers one inch and a quarter in diameter. The sepals and petals are yellowish green with fine purple dots, the former three, the latter two lines broad. Lip white, half an inch long, striped with violet in the middle.

596. MAHARANGA EMODI. *(aliàs* Onosma Emodi *Wallich.)* A hardy perennial from Nepal. Flowers pale rose-colour, small. Belongs to Borageworts. Introduced by Major Madden. (Fig. 291.)

Our knowledge of this singular herbaceous plant is owing to Mr. Moore, the Superintendent of the Botanic Garden, Glasnevin, who forwarded us specimens in May, 1851. The seeds were sent him from the Himalayas, by Major Madden, under the name of *Onosma Emodi,* and it corresponds with specimens so marked in our herbarium, for which we are indebted to Dr. Wallich. M. Alphonse De Candolle seems, however, to intend the present plant by his *Maharanga Wallichiana,* at least it agrees with its specific character in the *Prodromus,* and not with that of his *M. Emodi.* We have not indeed seen any plant to which the latter definition will apply, none of our Maharangas having from three to five-nerved leaves. The genus is distinguished from Onosma by the presence of a plaited coronet inside the tube of the corolla above its insertion, and by the peculiar form of that organ, which consists of a short cylindrical tube expanding suddenly into a great ovate closed limb. The name Maharanga is that employed by the Nepalese, among whom the great fusiform root of the plant is used in producing a blue dye : it is said to signify " a strong or intense colour." Dr. Wallich, in Carey's *Flora indica*, thus describes the species as it occurs in Nepal :—

"Root stout, sub-fusiform, dividing at the end into several thick branches, whitish within, covered with deep purple bark ; fibres capillary, few. Stem slender, round, divided into simple branches, as well as all the other parts, covered with small circular dots, each terminating in a straight, simple bristle. Leaves scattered, sessile, hispid and dotted above, smoother below, with three longitudinal nerves, uniting a little above the base, sometimes with another pair from the middle rib, varying considerably in size, mostly lanceolate, four or five inches long ; sometimes sub-linear, and in that case generally shorter. Racemes gradually expanding and becoming erect as the flowers open, very hispid, one or two inches long. Flowers small, copious, secund, erect, on short pedicels, which equal their linear, solitary bracts. Calyx ovate, five-angled, growing larger with the ripening seeds ; laciniæ triangular, acute, the base of their sinuses forming five prominent corners. Corolla pale, bluish towards its mouth, twice the length of the calyx, hairy, five-keeled, with as many deep furrows ; the base inverted over the ovaria, and embracing the base of the style ; throat contracted ; laciniæ ovate, acute. Filaments inserted on five villous protuberances, below the middle of the corolla, corresponding to the external five furrows ; anthers linear-sagittate, larger than the filaments, converging into a cone ; their slightly twisted bases cohering. Style longer than the corolla, slender ; stigma annular. Seed brownish, shining, dotted, and tubercled, keeled on the inner side, ending in a compressed short beak, and in other respects exactly like those of *O. simplex*, Gaert. Carp., i. 325, tab. 67.

"*Obs.* I should have taken this plant to be the same as *O. tinctoria*, had any of the authors I have consulted, and who appear to have copied Marschal a made any allusion to the remarkable inwards, forming a narrow margin, of the pistil ; its middle is sharply five-large protuberances on which the stamina material for dyeing blue, and imported Thibet, as a drug, under the native name, Bieberstein's description of that species, structure of the corolla. Its base is bent which closely embraces the lower part keeled, and marked within with five are inserted. The root is used as a from Gosain Than, probably also from mentioned above."

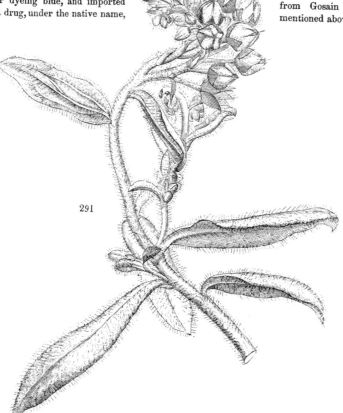

291

PLATE 97.

L.Constans del.& zinc.

Printed by C.F.Cheffins,London.

[Plate 97.]

THE AZOREAN FORGET-ME-NOT.

(MYOSOTIS AZORICA.)

———◆———

A brilliant half-hardy Perennial, from the Azores, *belonging to* Borageworts.

═══════════════

Specific Character.

THE AZOREAN FORGET-ME-NOT. Stem decumbent, much branched, covered all over with close bristly reflexed hairs. Leaves spreading, the hairs on the upper side close pressed, on the under side turned backwards ; the lower oblong-spathulate, the upper oblong and obtuse. Racemes dense, without bracts, forming corymbs when flowering. Calyx nearly five-parted, as long as the erect footstalk, covered with close-pressed hairs, eventually spreading, and as long as the tube of the corolla. Nuts very smooth.

MYOSOTIS *AZORICA ;* caule basi decumbente ramosissimo ubique densè setoso-hirsuto pilis reflexis, foliis patentibus pilis supernè adpressis subtus retrorsis hirsutis, inferioribus oblongo-spathulatis superioribus oblongis obtusiusculis, racemis ebracteatis densifloris sub anthesi corymbosis, calycibus sub-5-partitis pedicello erecto æqualibus adpressè aut subadpressè pilosis demum apertis longitudine tubi corollæ, nuculis lævissimis.—*De Cand. Prodr.,* 10. 106.

Myosotis azorica : *H. C. Watson, in Bot. Mag.,* t. 4122.

═══════════════

"This beautiful Forget-me-not is found about waterfalls, and on wet rocks with a north-east aspect, in the islands of Corvo and Flores, the most westerly of the Azores. Its proper habitat appears to be on the mountains ; though it comes down nearly to the sea-shore, following the course of rocky mountain streams, where the atmosphere is kept humid by the spray of the water. The deep rich blue of its numerous flowers, and their long succession from the lateral branches, combine to render this species well deserving of cultivation, provided it can be brought to flourish in the drier climate of our gardens. It will require a loose, peaty, or sandy soil, careful shading from the midday sun, frequent sprinkling with water, and to be covered with a glass in hot dry weather. Under this treatment a plant of it in my garden has completely filled with its numerous stems a square hand-glass, twenty inches to the side, and twenty-four inches in depth; and apparently it would have grown larger, had space allowed the free development of the lateral branches, which are much cramped by the glass. It will bear some frost, but may likely prove more impatient of cold than our native species of the genus. In a Wardian case it would probably succeed very well."

Such is the account given of this charming plant by Mr. Hewitt Watson, its discoverer. We find it thrive perfectly well in a greenhouse, among Heliotropes and Pelargoniums, where it ripens its little black glossy nuts (seeds) in tolerable abundance. The play of colour in the many-tinted flowers and flower-buds is scarcely rivalled by anything in cultivation.

PLATE 98

L.Constans del.& zinc.

Printed by C.F.Cheffins, London.

[PLATE 98.]

THE DUKE OF DEVONSHIRE'S WATER LILY.

(NYMPHÆA DEVONIENSIS.)

———◆———

A very brilliant HYBRID AQUATIC, *with Crimson Flowers.*

Nymphæa Devoniensis : *Paxton, in Gardener's Chronicle, July 10, 1852 ; Hooker, in Botanical Magazine,* t. 4665.

"How is it that aquatic plants are seldom or never brought under the influence of hybridism? They are objects of great beauty, are and always must be much in request, and appear to be just as submissive to man as other plants. Their constitutions may certainly be affected by crossing, quite as much as a Rhododendron. Yet, while the tender crimson species of Indian Rhododendron are brought to act upon the hardy pale faces of the United States, the delicate white Water Lily of our rivers is left to wild nature in the presence of the most glowing tints possessed by her tropical kindred.

"It may be said that there are physical difficulties in the way of crossing Water Lilies. We grant it. The yellow Nuphars are not likely to breed with the white and blue and crimson Nymphæas, and perhaps Victoria may refuse all alliance with either. But then it is the same everywhere; a Currant will not breed with a Gooseberry, nor an Apple with a Pear. Nevertheless, Gooseberries find kindred blood among Gooseberries, and Currants among Currants : and why may it not also happen to the Nymphæas themselves? This sort of crossing is certainly possible. It has been done.

"Some years since mules were obtained in the Horticultural Garden between the tender blue Nymphæa of the Cape of Good Hope and the hardy white one of England. But owing to neglect they were allowed to perish, and that experiment came to nothing.

"At this moment there is actually flowering at Chatsworth a mule produced by crossing *Nymphæa rubra* with *N. Lotus.*

"Seeds were obtained in the autumn of 1850, and from them in the following summer Sir Joseph Paxton had the gratification of finding himself in the possession of a most beautiful hybrid, which he named Devoniensis, after the duke, his patron. In leaf and flower it has a great advantage in point of size and robustness of growth over either of its parents; but its most valuable property is its continuing to flower the whole of the season without intermission. The parent plant produced its first flower as early as the 12th of April, 1851, and continued to flower until the middle of October, when it was removed, with a fine succession of flower-buds still upon the plant, to its winter quarters. During this period it often had two expanded flowers and five buds in different stages of development. It produces its flowers quite as freely as *N. dentata*; and its beautiful colour (which is not quite so deep as its parent), together with its large size which has often been as much as eight inches in diameter, together with its fine leaves which have been seldom less than thirteen to seventeen inches across, renders it one of the best Nymphæas in cultivation.

"Let us hope that this example will not be thrown away. The season has come; the Nymphæas are all in flower, or nearly so; and there can be no difficulty in operating to any extent upon the white Nymphæa, which we should take for the mother of the brood that it is hoped will come."

The plant thus referred to in the *Gardener's Chronicle* is now represented from a specimen received from Chatsworth, and it will be admitted that it deserves all that was said of it. It has also been published in the *Botanical Magazine* by Sir W. Hooker, who states that for the opportunity of figuring this truly splendid plant, he is indebted " to Mrs. Spode, the lady of Joshua Spode, Esq., Armitage Park, Rugely, Staffordshire, whose gardens and rare exotics are celebrated in the neighbourhood, and are likely to be still more so from the taste and skill displayed by their generous proprietors, and by the zeal and energy of their intelligent head gardener." Sir William adds that the living plant at Kew, from Mrs. Spode, as well as cut specimens received from Armitage, and others sent by Mr. Davison from Sir W. Molesworth's tropical aquarium at Pencarrow, Cornwall, amply justify all that is said in the *Gardener's Chronicle*.

Mr. Davison observes, that with him *Devoniensis* grows and flowers most freely, planted in rough turf taken from a pasture and laid in a heap one year previous to its being used, with one-sixth of dried cow's-dung. The water in the tank in which it grows is kept from 75° to 80°.

We should add that Sir W. Hooker raises the question of whether *N. dentata* may not have been one of the parents of *N. Devoniensis*, rather than *N. Lotus*. He remarks that *N. Lotus* and *N. dentata* are very closely allied species, if they be really and truly distinct. He thinks that the pale and depressed base of the calyx of *N. dentata*, giving that part a somewhat conical form, furnishes what may perhaps prove a distinguishing mark, and that character he finds in *N. Devoniensis*. Mr. Davison, at Pencarrow Gardens, also speaks of the *N. Devoniensis* as "a hybrid between *N. rubra* and *N. dentata*." We have no means of assisting in this enquiry.

PLATE 99.

L.Constans del.& zinc.

Printed by C.F.Cheffins, London.

[PLATE 99.]

THE THICK-LEAVED CLEISOSTOME.

(CLEISOSTOMA CRASSIFOLIUM.)

———◆———

A very pretty Hothouse Epiphyte, from the EAST INDIES, *belonging to the Natural Order of* ORCHIDS.

Specific Character.

THE THICK-LEAVED CLEISOSTOME. Leaves fleshy, channelled, curved, stiff. Panicle simple, with the branches closely spicate and nodding. Lip with the lateral lobes erect and very small, the middle one roundish, with a small recurved tooth on either side. Tooth of the spur blunt and fleshy.

CLEISOSTOMA *CRASSIFOLIUM ;* foliis carnosis canaliculatis arcuatis rigidis, paniculæ simplicis ramis densè spicatis nutantibus, labelli lobis lateralibus minutis erectis intermedio subrotundo dente utrinque runcinato, calcaris dente parvo obtuso carnoso.

A VERY distinct species of Cleisostome, imported from some part of the East Indies, probably Moulmein, by Messrs. Veitch and Co. It is remarkable for its thick tough aloe-like leaves, and panicles of dense sea-green flowers, singularly enlivened by a rose or violet lip. The inflorescence, too, although, as is customary among Cleisostomes, consisting of small flowers collected into dense spikes at the end of the branches, has a peculiar curved or drooping appearance, by which the species may be known irrespective of its foliage.

Sepals oval, blunt, nearly equal, spreading. Petals with a similar form and the same direction, but very much smaller. Lip with a blunt oblong spur, filled with honey, one-celled, and twice as long as the limb, of which the lateral lobes are very short and erect, and the middle one rounded, with a minute tooth near the base on each side, while the point is so much reflexed as to be hidden unless the lip is lifted up. At the base of the column stands the characteristic tooth in the form of a blunt fleshy process, partly closing up the entrance to the spur. The pollen-masses are four, very small, pear-shaped and distinct, at the end of a filiform caudicle attached to an oblong gland.   In this respect the plant is at variance with other Cleisostomes, such species as we have examined having the pollen-masses in pairs, the lobes of which are unequal and plano-convex.

We observe that the late Mr. Griffith enquires in his *Notulæ* (p. 358) why Cleisostoma is separated from Saccolabium and Sarcanthus.   The differences among the three genera are these :— In Saccolabium the spur of the lip is one-celled, without any tooth at the foot of the column ; to Cleisostoma and Sarcanthus that peculiar process is essential.   In Cleisostoma the spur is absolutely one-celled, while in Sarcanthus it is more or less completely two-celled.   It is a question, no doubt, whether Blume's genus Cleisostoma ought to be separated from Sarcanthus, but about the distinctness of Saccolabium we entertain no doubt.

GLEANINGS AND ORIGINAL MEMORANDA.

597. GAURA LINDHEIMERI. *Engelmann.* A hardy perennial, with white and pink flowers. Native of Texas. Belongs to the Order of Onagrads. (Fig. 292.)

A branching herbaceous plant, growing from three to four feet high, and producing an abundance of gay white and reddish flowers during all the latter part of the year. The branches are long, rod-like, naked except at the extremities where the flowers grow. The lower leaves are deeply divided in a pinnatifid or sinuate manner; the upper are lanceolate and slightly toothed, the uppermost of all are linear-lanceolate and entire. The flowers appear in long virgate spikes, which frequently branch near the end. The petals are pure white; the flower-buds are long and slender, green when young, a warm reddish brown just before expansion. The seed-vessels are small sessile four-cornered nuts. A perennial, growing freely in any good garden soil, and flowering from July to September. It is easily increased from seeds, and is best treated as a half-hardy biennial. It will not flower before the second season. Having been found in Texas or provinces more to the southward, it cannot be regarded as perfectly hardy. It is really a showy although a straggling plant, and well suited for decorating mixed beds of flowers, or the skirts of a plantation in the autumn.—*Journ. of Hort. Soc.,* vol vii.

292

598. GOETHEA STRICTIFLORA. *Hooker.* An uninteresting hothouse shrub from Brazil, belonging to the Natural Order of Malvads. Flowers whitish.

A very remarkable-looking plant, sent to us by Messrs. Rollison, and by Mr. Henderson, St. John's Wood, under the name of *Goethea cauliflora* of Nees von Esenbeck. But it is certain that the plant can neither be the *G. cauliflora* of Nees and Martius, nor his *G. semperflorens.* Our plant has the leaves broad-ovate and sinuato-dentate, and the flowers invariably erect from the axils of the leaves. The flowers are very inconspicuous, and quite concealed by the involucre, whose beautiful red-veined bracts, looking like a calyx, persist long after the blossoms have passed. Leaves alternate, large, petiolate, ovate, often broadly so, acuminate, penninerved (with three principal nerves from near the base), the upper half sinuato-dentate at the margin. Peduncles short, aggregated in the axils of the leaves (and often remaining after the leaves are fallen, above the scars), scarcely half an inch long. Involucre of four erect, pale, yellowish-white, cordate bracteas, striated and veined with red, including a single flower, whose stigmas alone are sometimes protruded beyond the involucre. Calyx nearly white, or greenish, cut into five erecto-connivent acuminated lobes. Corolla of five obcordate, veiny, small petals, which are united by their base to the cylindrical tube of the filaments of the anthers, shorter than the calyx. Style as long as the tube of the anthers, then separating into ten branches, each bearing a capitate stigma.—*Bot. Mag.*, t. 4677.

599. CERASUS LAUROCERASUS; *var.* Pumilio.

This is a curious dwarf variety, resembling the common Laurel in much the same way as the Clanbrazil Fir resembles a Spruce. The leaves are from two to three inches long, and the habit extremely dwarf. If it does not hereafter run away, it will be a useful variety for places where the common Laurel is too large. A plant was received by the Horticultural Society in 1851, from Lieut-General Monckton, F.H.S., whose brother's gardener, William Reynolds, raised it from seed of the common Laurel.—*Journ. of Hort. Soc.*, vol. vii.

600. HELIOPHILA PILOSA. *Lamarck; var.* arabidoides *Sims.* A hardy annual, native of the Cape of Good Hope. Flowers bright blue. Belongs to the Cruciferous Order. (Fig. 293.)

293

This little grown plant deserves to be reintroduced to cultivation, for its flowers are of the most brilliant blue, and although fugitive are so incessantly renewed, that the effect of a bed of it is nearly as good as that of a blue Lobelia. The late Mrs. Wray used to grow it charmingly, as a hardy annual, raised on her vine borders at Cheltenham. It is an annual, native of the Cape of Good Hope, whence it was long ago introduced, and then received the name of *H. arabidoides;* but De Candolle regarded it as a mere variety of *H. pilosa,* which is probable enough, for the cultivated plant varies much in the quantity of hairs that it produces; sometimes, in wild specimens it is almost shaggy; at other times, in cultivation it is, so nearly smooth that our artist overlooked the few that continue to appear. It grows about eighteen inches high and ripens seed plentifully. The long narrow pods are uniformly dilated at the end, as if attempting to assume the necklace form observable in so many species of the genus; and the pair of short filaments is always furnished with a conspicuous dorsal tooth. Our cut has been made from specimens communicated many years ago by Mrs. Wray.

601. PELARGONIUM FOLIOLOSUM. *De Candolle.* (*aliàs* Geranium pinnatum *Andrews.*) A tuberous-rooted greenhouse plant with pale yellow flowers. Native of the Cape of Good Hope.

This was purchased from Mr. Wicks, a collector of Cape plants, May 3rd, 1852, as a Yellow Pelargonium. It is one of the fleshy-rooted species, often called Hoareas. It has hairy pinnated leaves, with about seven pairs of ovate entire leaflets, and an odd one, which is much broader and rounder. The flower-stem grows higher than the leaves, and divides into two unequal arms about the middle; of these, one flowers some weeks before the other. The umbels consist of six or eight blossoms, with hairy stalks three times as long as the subulate bracts. The petals are linear, channelled, recurved, blunt, pale clear buff, the two upper standing nearer to each other, and with a deep crimson spot in the middle. This was obtained for the sake of its yellow flowers, which it is hoped may be made to change the colour of some of the large-flowered Pelargoniums. As the pollen is good, this may happen. It requires a good rich sandy soil, and to be treated like the ordinary kinds of Pelargoniums; but it must be kept rather dry in winter. As has been stated, its value will be as a breeder; the flowers are too insignificant to render it of importance otherwise in a gardening point of view.—*Journ. of Hort. Soc.*, vol. vii.

602. STANHOPEA BUCEPHALUS. *Lindley; var.* guttata. A beautiful stove epiphyte, with deep orange spotted flowers. Blossoms in September.

This very fine variety has been sent us by Mr. James Napier, gardener at Corehouse near Lanark. We are unacquainted with its native country. Fourteen flowers appeared upon a single spike, the largest number yet remarked in any Stanhopea. The lip has the peculiar long narrow hypochil and short smooth mesochil which so distinctly characterise the original species; but the sepals, petals, and hypochil are a deep apricot orange-colour; on the hypochil are four deep brown blotches, two outside and two inside; the sepals have no spots; on each petal there are four, two at the base, and two above the middle, so that there are in all twelve broad brown stains; the epichil is brightly speckled, but at the base only.

603. HEINTZIA TIGRINA, *Karsten.* A magnificent hothouse plant, belonging to Gesnerads. Flowers rose-colour and white. Native of the Caraccas. (Fig. 294.)

This appears, from the *Flore des Serres*, to have found its way into the gardens of Germany. It is one of the noblest plants of its noble race. The leaves are often a foot long, black green with purple ribs on the paler under side. The calyx is rich rose-colour, with a green rib in the middle of each sepal. The corolla is pure white, with blood-coloured spots on the limb or expanded part. Mr. Karsten, its discoverer, says it grows five or six feet high in shaded places on the mountains of the Caraccas, at an elevation of 5000 feet, among Ferns of various kinds, where it flowers in the summer months. It must be grown in such a high temperature and moist atmosphere as suit the more tender plants of the order, Sinningias, Gloxinias, and Nematanths.

604. BRYA EBENUS. *De Candolle.* (*aliàs* Amerimnum Ebenus *Swartz ; aliàs* Pterocarpus glabra *Reichard ; aliàs* Pterocarpus buxifolius *Murray ; aliàs* Brya arborescens *Browne ; aliàs* Aspalathus arboreus, &c. *Sloane.*) A stove shrub with bright yellow flowers. Belongs to the Leguminous Order. Native of the West Indies, where it is called *Jamaica Ebony.*

A well-known West Indian shrub, or rather tree, especially common in Jamaica, whence our plant was derived ; but it is little seen in cultivation, by no means so much as it deserves ; for although in its native country it attains a

294

height of fifteen or twenty feet (M'Fadyen; Sloane says forty feet), yet, cultivated in a pot, in a warm stove, it maintains a shrubby character for a very great number of years, with pretty, evergreen, box-like foliage, bearing copious bright orange pea-shaped flowers in the month of May, yielding a delicious perfume. It abounds in the savannas and dry hills of Jamaica, where, Dr. M'Fadyen says, with its long twiggy branches, it reminds the traveller of the Broom of Europe. The wood is hard and ponderous, of a fine greenish brown colour, susceptible of a good polish, and used formerly to be imported into Europe; but it is extremely different from the true Ebony of commerce, *Diospyros-Ebenus* of Madagascar; and the trunk, rarely exceeding four inches in diameter, can only yield small samples for cabinet-work. " The slender branches," says Patrick Browne, " are very tough and flexile, frequently used for riding-switches, and in his days (days happily now gone by) generally kept at all the wharfs about Kingston to scourge the refractory slaves." A shrub or small tree, from eight or ten to forty feet high, with long twiggy branches, armed with short, sharp, subulate, stipulary spines. Leaves solitary or in clusters, box-like, evergreen, obovato-cuneate, sessile. Flowers axillary, solitary, or two or three together. Peduncle short, with a pair of minute, opposite, small bracts above or near the middle. Calyx bell-shaped, pubescent, obscurely two-lipped; upper lip bipartite, lower tripartite; segments ovate, acute, the lowest one spreading, the rest erect. Corolla bright orange-yellow. Vexillum subrotund, with deep purple streaks in the centre. Alæ and carina oblong, somewhat falcate, obtuse: all the petals with short claws. Stamens ten, monadelphous, nearly as long as the alæ. Anthers subglobose. Pistil hairy. Ovary oblong, of two joints, the upper side with an even line, below bigibbose, the upper joint tapering into a long subulate style; stigma a mere point. " Legumen pedicelled, not an inch in length, compresso-foliaceous, with the valves chartaceous, hirsute with minutely capitate hairs, biarticulate; lower joint with the upper suture nearly straight, and the under convex; upper joint small, abortive." *M'Fadyen.—Bot. Mag.*, t. 4670.

295

605. OPHIOXYLON MAJUS. *Hasskarl.* (*aliàs* O. album *Siebold.*) A neat hothouse shrub, native of the East Indies. Belongs to Dogbanes. Flowers white in April. (Fig. 295).

There has been a difference of opinion among botanists whether there are one or two species of Ophioxylon; but the question would seem to be set at rest by Mr. Hasskarl, who describes this plant as being altogether stronger in growth, with a smooth shrubby stem four feet high, leaves green beneath, white flowers, and olive-shaped fruit, while in *O. serpentinum* this plant does not grow above a foot high, is not a shrub, has leaves red underneath, larger reddish flowers, and globose fruit. The large white-flowered plant, O. *majus*, thought by Hasskarl to be possibly the *Ophioxylon album* of Gærtner, forms in the stove a small light green shrub with oblong-lanceolate membranous leaves placed in threes or fours, and loose cymes of white flowers. The corolla is nearly three quarters of an inch long, with the lobes of the limb half circular. It grows freely in a mixture of sandy loam and peat; but requires to be kept in rather a moist atmosphere. It is increased by cuttings put in sand under a bell-glass, and plunged in the bark bed. The plant is of little value in a horticultural view, the white flowers being too small to produce a striking effect. It is however of some medical interest, being one of the plants whose roots are believed by Indian practitioners to be a cure for the bite of venomous serpents.—*Journ. of Hort. Soc.*, vol. vii.

606. SALVIA RŒMERIANA. *Scheele.* (Linnæa, xxii. 586.) A pretty sub-shrubby half-hardy plant, with spikes of crimson flowers, produced all the summer. Native of Texas, " in woods near Neubraunfels." Belongs to Labiates. Flowered in the Chelsea Botanic Garden.

Stems two feet high, branched, villous, quadrangular. Leaves on longish hairy petioles, which are dilated and somewhat connate at the base, and slightly furrowed above; rugose with coarse sunken reticulated veins which are prominent beneath, pilose on both surfaces, with numerous sessile glands, (which are flame-coloured when dry); dark dull green above, paler and grayish beneath. The lower leaves are irregularly pinnately cut, in the cultivated plants producing a pair of very small obovate subopposite leaflets, and a many times larger broadly ovate almost reniform terminal one, which is cordate at the base, and deeply and irregularly crenate-lobed on the margin. The upper leaves are simple cordate-ovate, deeply crenate-lobed. The inflorescence forms spikes of eight to ten inches long, with distant three to four-flowered verticillasters, in the axils of oblong-lanceolate villous bracts equalling the peduncles. The flowers are small, of a very rich crimson. Calyx green, thirteen-nerved, turbinate-campanulate, two-lipped, clothed with white hairs inter-mixed with glands (flame-coloured when dry); upper lip truncate, the teeth connivent, the two lateral teeth shortly cuspidate; lower lip of two ovate-lanceolate sharp-pointed teeth nearly as long as the tube. Corolla tubulose, much exserted, nearly three times as long as the calyx, the tube enlarged above, puberulous outside, and with a broad ring of hairs within near the base; upper lip erect, concave, emarginate; lower lip patent, trifid, the lateral lobes rounded, ovate, spreading, the middle lobe transverse, broader, emarginate, style and filaments red, the cells of the anthers separated by the prolonged connective; stigma bifid, the lobes recurved.—*T. Moore.*

607. CAMPANULA VIDALII. *Watson.* A half-hardy undershrub. Flowers large, dirty white. Native of the Azores. (Fig. 296.)

This species was first made known through Sir William Hooker's *Icones*, by Mr. Hewitt Watson, to whom it was given by Captain Vidal, R.N., whose name it bears. It was found on an " insulated rock off the east coast of Flores,

296

between Santa Cruz and Ponta Delgada." Seeds were received some time since from Mr. Ayres, who was indebted for them to Mr. P. Wallace. The plant has a fine handsome deep green shining succulent foliage, and forms a very good-looking decumbent shrub. Some of the shoots are merely terminated by long rosettes of leaves; others throw up an erect, graceful, flowering stem, with a shiny surface, and a warm greenish-brown colour, terminated by several large white nodding flowers, each about an inch and a half long, and shining as if glazed. The colour is, how-ever, bad, a tint of dull purple or even pale cinnamon giving them a dirty ap-pearance. It is a half-hardy or green-house shrub, growing best in a mixture of sandy loam and leaf-mould, increasing freely by seeds, but not flowering before the second season from seed. It blossoms in August, and is a good object for rock-work in a climate which suits it; but, being tender, its value is much diminished, independently of the dingy colour of its flowers. This plant has so little the ap-pearance of an ordinary Campanula that it is a question whether it truly belongs to the genus. It would rather seem to be related to *Musschia,* the old *Campanula aurea,* though by no means to be asso-ciated with it. The ovary is three-celled, with a great rugged double placenta ex-panding in each cavity, and around the flat head of the ovary, inside the corolla, there runs a broad yellow fleshy ring-like disk; but neither in this nor in any other circumstance, except habit, does there appear to be real ground for generic separation.—*Journ. of Hort. Soc.,* vol. vii.

608. IMPATIENS MACROPHYLLA. *Gardner.* A hothouse perennial, with small orange-coloured flowers. Native of Ceylon. Belongs to the Natural Order of Balsams. Introduced at Kew.

We have here another of the many curious species of Balsam which abound so much in Ceylon, and we may say perhaps in the moist and mountainous parts of India generally. Our gardens are indebted for seeds of this to Mr. Thwaites, the able superintendent of the Botanic Garden at Peradenia, who sends it to us from Adam's Peak (No. 436 of Mr. Thwaites' dried collection), and Mr. Gardner's specimens (No. 159 of his collection) are from Newra Ellia, at 6000 feet of elevation. We had, many years ago, received Ceylon specimens, without any particular locality, from Mrs. General Walker. Our plants flowered at the Royal Gardens, in a moist but not very hot stove, in the early summer of the year after the seeds were sown ; and, small though the blossoms are, yet their deep tawny orange-colour, stained with red, and the numerous long bright petioles, together with the ample foliage, render this a handsome plant. Our plants attain a height of from two to three feet ; in their native country they are probably much taller. The stem is erect, straight, as thick as, or thicker than, one's finger, purplish. Leaves mostly at the top of the stem, below them are the scars of many fallen ones ; they are crowded, alternate or scattered, large, five to six inches long (some of our native specimens measure nearly a foot), ovate, much and gradually acuminated, pilose on both sides, dark green above, paler beneath, closely penninerved ; the margin everywhere serrated, the serratures mucronate ; at the base the margin is fringed with long soft bristles, tipped with a gland, and is gradually attenuated into the long, stout, bright, red leaf-stalk upon which are a few scattered glandular setæ. Peduncles axillary, aggregated (often densely crowded), much shorter than the petioles, single-flowered, having minute bracteas at the base. Flowers small for the size of the plant, deep tawny-orange, stained with red. The upper sepal is oblong, convex, red, terminated with a long claw-like point. The lower one, or labellum, is cucullate, the mouth ending in a sharp recurved acuminated point, like the mouth of a ewer : the spur is short, hispid, with a few long bristles, singularly incurved almost upon itself, and swollen and didymous at the apex.—*Bot. Mag.,* t. 4662.

297

609. SEDUM PURPUREUM. *Link.* (*aliàs* S. purpurascens *Hort.*) A hardy herbaceous plant, with purple leaves and flowers. Native of Russia. Belongs to the Order of Houseleeks. (Fig. 297.)

By many writers this is regarded as a mere variety of *Sedum Telephium,* and their opinion is probably correct. It only differs in being pervaded by a very deep purple tint, and in the leaves being wedge-shaped and narrow at the base, instead of being oblong and rounded at the base. The petals also are flat, not channelled at the point, and the stamens are rather longer than the petals. It grows naturally in middle Russia, and all over Siberia, whether in the Altai, the Ural, or the Baical, reaching even to Kamtchatka. In cultivation it is a hardy plant, growing eighteen inches in height in any good light rich soil. It is increased by dividing the old plant in the ordinary way. It flowers in August. It is a rather showy and desirable plant for rock-work in summer.—*Journ. of Hort. Soc.,* vol. vii.

610. RESTREPIA NUDA. *Klotzsch.* A stove epiphyte, belonging to Orchids. Native of Venezuela. Flowers white. Introduced by M. Allardt of Berlin.

Restrepia *nuda ;* caulibus secundariis cæspitosis, basi vaginatis teretibus ; foliis carnosis solitariis acutis planis versus basin attenuatis ; floribus pedunculatis paucis nudis ; perigonii foliolis 2 candidis rubro striatis elongato-lanceolatis acuminatis, supremo trinervio, inferiore 4-nervio, interioribus basi lanceolato-dilatatis candidis, margine denticulatis, dorso acumineque setiformi purpureis ; labello purpureo elongato-obovato acuminato margine fimbriato ima basi auriculato ; gynostemio clavato.

Stems two to three inches long, cæspitose. Leaf leathery, shining, three to four inches long. Flowers solitary, one inch and a half long. Sepals white striped with red, an inch long, three to four lines broad ; petals ten lines long.—*Allgem. Gartenzeit., Aug.* 28, 1852. The pollen-masses not being mentioned, it is uncertain whether this is a Restrepia or a Pleurothallis.

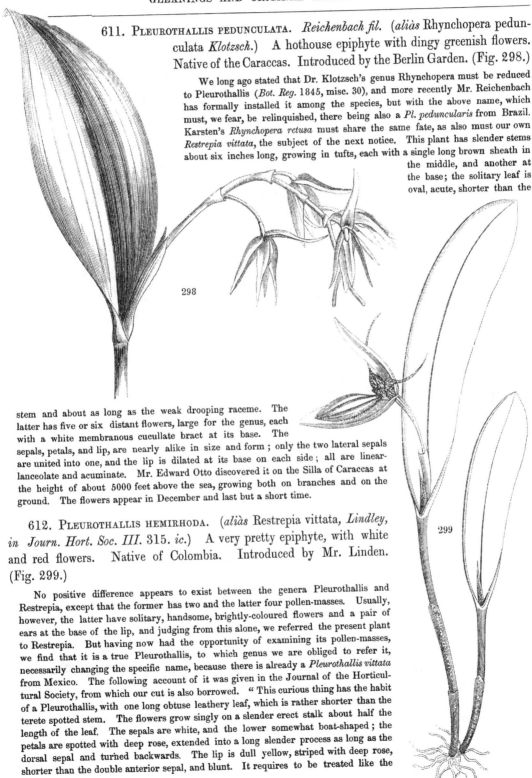

611. PLEUROTHALLIS PEDUNCULATA. *Reichenbach fil.* (*aliàs* Rhynchopera pedunculata *Klotzsch.*)   A hothouse epiphyte with dingy greenish flowers. Native of the Caraccas. Introduced by the Berlin Garden. (Fig. 298.)

We long ago stated that Dr. Klotzsch's genus Rhynchopera must be reduced to Pleurothallis (*Bot. Reg.* 1845, misc. 30), and more recently Mr. Reichenbach has formally installed it among the species, but with the above name, which must, we fear, be relinquished, there being also a *Pl. peduncularis* from Brazil. Karsten's *Rhynchopera retusa* must share the same fate, as also must our own *Restrepia vittata*, the subject of the next notice. This plant has slender stems about six inches long, growing in tufts, each with a single long brown sheath in the middle, and another at the base; the solitary leaf is oval, acute, shorter than the stem and about as long as the weak drooping raceme. The latter has five or six distant flowers, large for the genus, each with a white membranous cucullate bract at its base. The sepals, petals, and lip, are nearly alike in size and form ; only the two lateral sepals are united into one, and the lip is dilated at its base on each side ; all are linear-lanceolate and acuminate.   Mr. Edward Otto discovered it on the Silla of Caraccas at the height of about 5000 feet above the sea, growing both on branches and on the ground.   The flowers appear in December and last but a short time.

612. PLEUROTHALLIS HEMIRHODA. (*aliàs* Restrepia vittata, *Lindley, in Journ. Hort. Soc. III.* 315. *ic.*) A very pretty epiphyte, with white and red flowers.   Native of Colombia.   Introduced by Mr. Linden. (Fig. 299.)

No positive difference appears to exist between the genera Pleurothallis and Restrepia, except that the former has two and the latter four pollen-masses. Usually, however, the latter have solitary, handsome, brightly-coloured flowers and a pair of ears at the base of the lip, and judging from this alone, we referred the present plant to Restrepia. But having now had the opportunity of examining its pollen-masses, we find that it is a true Pleurothallis, to which genus we are obliged to refer it, necessarily changing the specific name, because there is already a *Pleurothallis vittata* from Mexico. The following account of it was given in the Journal of the Horticultural Society, from which our cut is also borrowed. " This curious thing has the habit of a Pleurothallis, with one long obtuse leathery leaf, which is rather shorter than the terete spotted stem. The flowers grow singly on a slender erect stalk about half the length of the leaf. The sepals are white, and the lower somewhat boat-shaped ; the petals are spotted with deep rose, extended into a long slender process as long as the dorsal sepal and turned backwards. The lip is dull yellow, striped with deep rose, shorter than the double anterior sepal, and blunt. It requires to be treated like the genus Pleurothallis, and is one of the handsomest of the race which that genus represents." The outer half of the flowers is pure white, the inner half more or less red : whence the name.

613. LILIUM GIGANTEUM. *Wallich.* (*aliàs* L. cordifolium *Don.*) A magnificent hardy bulbous plant from Nepal. Flowers white and fragrant, appearing in July.

The discovery of this Prince of Lilies we owe to Dr. Wallich, who detected it in moist shady places on Sheopore in Nepal. "This majestic Lily," he says, "grows sometimes to a size which is quite astonishing ; a fruit-bearing specimen of the whole plant, which is destined for the Museum of the Hon. East India Company, measures full ten feet from the base of the stem to its apex. The flowers are proportionably large and delightfully fragrant, not unlike those of the common white Lily." Nor does it degenerate in cultivation ; the flowering plant having attained a height of ten feet in one season ; the flower portion occupying twenty inches. Such a raceme of flowers, accompanied by leaves measuring ten to twelve inches long and eight inches broad, must have afforded a striking spectacle. Baron Hügel found the plant in the Peer Punjäl pass of the Himalaya, leading into Kashmeer ; and we believe that Drs. Thomson and Hooker met with it abundantly in other portions of that vast range of hills. The remainder of our account shall be taken from Dr. Balfour's notes, chiefly drawn up from the living plant at Comely Bank near Edinburgh. "Major Madden says the *Lilium giganteum* is common in the damp thick forests of the Himalaya, the provinces of Kamaon, Gurwhal, and Busehur, in all of which he has frequently met with it. It grows in rich black mould, the bulb close to the surface, at from 7500 to 9000 feet above the level of the sea, where it is covered with snow from November to April, or thereabouts. The hollow stems are commonly from six to nine feet high, and are used for musical pipes. The fruit ripens in November and December. Stem straight, cylindrical, smooth, gradually attenuated to the apex, nearly ten feet high, five and a half inches in circumference at the base, green with a reddish-purple hue at the upper part. Leaves alternate, scattered, the internodes varying in length, petiolate, broadly ovate, cordate, acuminate, shining dark green above, paler below, venation reticulated, having an evident midrib, with the veins coming off from it ending in an intra-marginal vein ; lower leaves with long petioles, very large, ten to twelve inches long, eight inches broad, becoming gradually smaller in ascending ; upper leaves small, sessile, ovate, acute. Petioles of lower leaves twelve to fourteen inches long, thick, broad and somewhat sheathing at the base, lower surface convex, upper with a deep and broad furrow ; petioles of upper leaves short. Bracts ovate, acute, caducous, leaving a semilunar scar. Flowers white, with purple sheaths, greenish below, infundibuliform-campanulate, inclined downwards, twelve on the raceme, fragrant ; tube greenish, two inches in circumference at the base, gradually dilating upwards ; limb slightly revolute ; leaves of the perianth oblong-spathulate, three outer with slight purple streaks inside, three inner rather broader, with a deep purple tinge on the inside, and with a prominent ridge on the outside, sulcated on either side, and two elevated ridges on the inner surface separated by a shallow groove."—*Bot. Mag.*, t. 4673. There is great reason to hope that this noble plant, of which Messrs. Veitch have raised an abundance, will prove hardy. At least it can require nothing more than a covering of ashes in winter.

614. VINCETOXICUM PURPURASCENS. *Morren and Decaisne.* (*aliàs* Cynanchum purpurascens *Siebold.*) A hardy herbaceous half-twining plant. Native of Japan. Belongs to the Order of Dogbanes. Flowers purple.

Stems and all the green parts slightly downy ; when in flower becoming weaker, with a tendency to twine. Leaves narrow, oblong, mucronate, becoming smaller near the ends of the shoots where the flowers appear. Flowers dull purple, on slender pedicels, in long-stalked many-flowered cymes, proceeding from the axils of the superior leaves, the size of, and very much like, the common *Vincetoxicum nigrum.* This perennial appears to be hardy, or half-hardy, like *V. japonicum,* growing with it freely in a peat border ; but, although transmitted as a good garden plant, it must be consigned to the mere botanical collector.—*Journ. of Hort. Soc.*, vol. vii.

615. PLEUROTHALLIS WAGENERIANA. *Klotzsch.* A stove epiphyte, of no great interest, belonging to Orchids. Native of Venezuela. Flowers yellowish. Introduced by M. Allardt of Berlin.

Pleurothallis (Aggregatæ) *Wageneriana ;* rhizomate funifero, squamis obtecto, caulibus secundariis 2—3 articulatis, vaginis 2 appressis obtusis subintegris obtectis ; folio crasso carnoso angusto primum conduplicato deinde canaliculato, basi cuneato apice attenuato retuso ; floribus binis brevi pedunculatis ; perigonii foliolis tribus exterioribus ringentibus crassis carnosis sordide flavidis, interioribus membranaceis sulcatis acutis flavidis striis parallelis purpureis notatis, exterioribus triplo minoribus ; labello trilobo atro-purpureo tumido, anticè tuberculoso scabro, perigonii foliolis interioribus æquantibus ; pedunculis bracteis hyalinis obtusis binis aut tribus subcucullatis vestitis.

Stem the thickness of a crowquill, three to four inches long. Leaf very thick and fleshy, the same length and half an inch broad. Flowers three lines long, with white bracts. Petals streaked with red and membranous. Lip deep red.—*Allgem. Gartenzeit.*, Aug. 28, 1852.

PLATE 100.

L.Constans del.& zinc.

Printed by C.F.Cheffins,London.

[PLATE 100.]

THE SCARLET SALPIGLOT.

(SALPIGLOSSIS COCCINEA.)

———◆———

A beautiful half-hardy annual, of GARDEN ORIGIN, *belonging to the Natural Order of* LINARIADS.

ALL we know of this beautiful novelty is that it has been raised near Colchester, and that it was sent to us last August by Messrs. Henderson of Pine Apple Place. It seems to differ from other Salpiglots in nothing except colour, which is here of a clear vivid tender scarlet, charmingly relieved by short veins of a deeper colour. As a garden plant it possesses high claims to distinction, for there are few annuals that equal it.

In a Botanical point of view it seems to confirm Mr. Bentham's opinion that all the so-called species of the genus, known by the names of *atropurpurea, straminea, picta* and *Barclayana* are mere forms of one wild but variable species, the *S. sinuata* of the *Flora Peruviana,* among which there is in reality no character available for specific distinction.

PLATE 101.

L.Constans del.& zinc.

Printed by C.F.Cheffins,London.

[PLATE 101.]

THE PRETTY RAPHISTEM.

(RAPHISTEMMA PULCHELLUM.)

◆

A fine stove climber, from the TROPICS OF ASIA, *belonging to the Order of* ASCLEPIADS.

Specific Character.

THE PRETTY RAPHISTEM. A twiner. Leaves heart-shaped, taper-pointed, membranous, smooth on each side, having glands on the upper side above the petiole. Segments of the corolla ovate, blunt, erect. Lobes of the coronet twice as long as the column. Stigma prominent, umbilicate.

RAPHISTEMMA *PULCHELLUM;* volubile, foliis cordatis acuminatis membranaceis utrinque glabris supra petiolum glanduliferis, corollæ laciniis ovatis obtusis erectis, coronæ stamineæ foliolis gynos egium duplò superantibus, stigmate prominulo umbilicato.– *Decaisne.*

Raphistemma pulchellum : *Wallich, Pl. As. rariores,* vol. ii. p. 50, t. 163 ; *Decaisne, in D.C. Prodr.,* viii. 516 ; *aliàs* Asclepias pulchella : *Roxb. Fl. Ind.,* ii. 54.

OUR knowledge of this fine new stove plant is derived from a specimen furnished last summer by Messrs. Weeks and Co., of the King's Road. Its large straw-coloured flowers, broad foliage, and twining habit make it a useful companion for the favourite Stephanotis; its leaves, however, are not so thick.

According to Dr. Roxburgh, " it is an extensive perennial twining species, native of the forests of Silhet, where it is called Kulum, flowering in the rainy season." To this Dr. Wallich adds Gualpara, Tavoy and Pegu; and that it is the largest flowered Asclepiad with which he is acquainted.

Dr. Roxburgh describes it thus : " Stems and branches twining; young shoots perfectly smooth and deep green. Leaves opposite, long-petioled, cordate, entire, smooth, acuminate, from four to eight inches long, and from three to six broad. Racemes very long-peduncled, sometimes proliferous; by age the rachis lengthens into the form of a short raceme. Flowers very large, pure white, long-peduncled. Calyx five-parted, smooth; corolla five-parted, rotate; segments oblong, in the bud imbricated. Nectary sub-cylindric: exterior lamina membranaceous, ensiform, ending in long, fine, acute points, which converge over the stigma, their texture horny and polished; in their retuse tops, are the pits where the anthers are lodged. Germs two, style short, common stigma five-angled; to the points of the angles the five ovate, hard, polished, chestnut-coloured bodies are attached, which give substantial support to the five pairs of large oval anthers, by means of their thick, short, polished, chestnut-coloured, cyathiform pedicels.—*Fl. Ind.* II. 55.

PLATE 102.

L.Constans del.& zinc.

Printed by C.F.Cheffins, London.

[PLATE 102.]

THE RACEMOSE SOLENID.

(SOLENIDIUM RACEMOSUM.)

———•———

A hothouse Epiphyte, from NEW GRENADA, *belonging to the Order of* ORCHIDS.

════════════

Generic and Specific Character.

SOLENIDIUM. Sepals equal, spreading flat, distinct. Petals of the same form. Lip unguiculate, bent downwards, with two elevated feathery plates which are free at the point, and have a keel between them at the base. Column straight, bordered with a membrane, one-toothed at the end on each side, with an elevated fleshy anther-bed; near the base on each side below the termination of the membranous border, is a gland. Pollen-masses two, waxy, excavated behind; caudicle linear; gland small and roundish. An epiphyte from tropical America, bearing pseudobulbs, and having the habit of Oncidium.

THE RACEMOSE SOLENID. Leaves two, narrowly strap-shaped, shorter than the racemose scape. Flower-stalks straggling. Lip linear, dilated and rounded at the point.

SOLENIDIUM; sepala æqualia, explanata, libera. Petala conformia. Labellum unguiculatum deflexum, lamellis 2 elevatis plumosis apice liberis, carinâ basilari interjectâ. Columna recta, membranaceo-marginata, apice utrinque 1-dentata, clinandrio elevato carnoso, basi utrinque infra alam glandulâ aucta. Pollinia 2, cereacea, posticè excavata; caudiculâ lineari, glandulâ parvâ subrotundâ. Herba epiphyta, Americæ tropicæ pseudobulbosa, Oncidii facie.

S. *RACEMOSUM;* foliis 2, angustè loratis scapo racemoso brevioribus, pedunculis divaricatis, labello lineari apice dilatato rotundato.

──────────────

Solenidium racemosum : *Lindley, in Orchidaceæ Lindenianæ,* no. 79.

════════════

A<small>N</small> epiphyte from the forests of New Grenada, near Pamplona, whence it was introduced by Mr. Linden; who states that it grows at the height of 8500 feet, flowering in November. For a fresh specimen we are indebted to Robert Hanbury, Esq., of Poles, with whom alone we believe that it has flowered.

140

The plant has much the appearance of an Oncidium, in its manner of growth, foliage, and flowers, but it is materially different in structure. The original definition of the genus, framed upon an examination of shrivelled and crushed flowers, is in some respects erroneous, and is now set right. The lip is not furnished near the end with two teeth; that appearance was produced by the two feathery plates which occupy the lip (fig. *b*) having been pressed into a mass inseparable from the lip itself; and the incumbent position of the pollen-masses with respect to their caudicle arose from the same cause and is not natural in the plant.

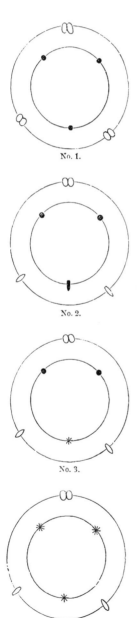

No. 1.

No. 2.

No. 3.

No. 4.

The main differences between Solenidium and Oncidium consist in this; that the column is earless and has a thin membranous border, terminating upwards in a thin triangular tooth, and rounded off above the base; beneath the lower end of the column stands a pair of distinct but minute glands, which must be analogous to the column ears of Oncidium, if there is any analogy between them. The crest of the lip, which in Oncidium is composed of three or some other uneven number of tubercles, is here replaced by a pair of long feathery plates which stand considerably above the lip itself, and being free at the end look in profile like a pair of shaggy ears. All this is very unsuccessfully represented on our plate at A. Variable as is the crest of the lip of Oncids it presents no structure approaching this, not even in the pulvinate division. The feathery plates are more like the raised lines of Cymbidium or Brassia, but the column and its peculiar basal glands resemble neither the one nor the other.

The feathery processes upon the lip, and the glands on the column, of Solenidium will be regarded as staminodes (abortive stamina), belonging—the first to the same series as the perfect stamen, and the last to a supposed inner series of undeveloped stamens, provided the theory referred to in *Folia Orchidacea* under Zygostates should be accepted by botanists. According to this theory the staminal apparatus of an Orchidaceous plant consists of two rings or whorls, each composed of three stamens more or less developed. In general the central of the outer whorl is alone perfect; while in Cypripedium perfection is confined to the two lateral inner stamens. The rest of the stamens are either wholly suppressed, as in many Dendrobes, or appear in the form of ears to the column or crests upon the lip; the ears of the column sometimes representing the lateral inner staminodes, and the crests of the lip being made up either of two lateral outer and one central inner staminode, or of either. Such evidence as exists upon this subject appears favourable to the opinion; which would be conclusively established if the crests of the lip were detected bearing pollen, a circumstance that has not yet been observed.

Upon this theory, the accompanying diagrams will represent the condition of the staminal apparatus in the different modifications which this Order produces. (In all cases but one, No. 5, the exterior ring represents the series to which the perfect stamen belongs, and the inner ring the series which is usually more or less disguised. For the convenience of description the perfect stamen and accompanying abortions may be called the *outer stamen and staminodes*, while those of the second and more paradoxical series may be termed the *inner stamen and staminodes*. The asterisks indicate an entire suppression of staminodes.)

No. 1 shows the theoretical state of the flower, with the three outer stamens complete, and three inner staminodes. The outer stamens are here in the condition in which they appear in the plant figured by Dr. Wight under the name of Euproboscis, and by Griffith in Falconer's Dendrobium normale.

No. 2 represents such genera as Odontoglossum in which one outer stamen is perfect, the two outer staminodes in the form of the lateral plates of the crest of the disk; then of the inner staminodes two form the wings of the column, and the other the midrib which separates or is blended with the lateral plates of the disk.

No. 3 represents such a structure as that of Anacamptis, where the usual outer stamen is attended by two of the inner staminodes, while two outer staminodes appear as plates on the lip, and the central of the inner staminodes is missing. Solenidium would also belong to this form.

No. 4 is the case of Cymbidium properly so called, in which all the inner staminodes are deficient, and the lateral outer staminodes lie upon the lip in the form of two raised lines.

No. 5 shows the beginning of the series in which outer lateral staminodes are wanting, except one which represents the perfect stamen in the preceding cases, while on the other hand the two lateral inner stamens are perfect and the third wanting; this occurs in Cypripedium.

No. 6. In Orchis the structure is absolutely reduced to one perfect outer stamen and a pair of inner lateral staminodes, occurring as tubercles at the base of the column; all the other staminal apparatus being missing. Thelymitra comes here.

No. 7 shows what happens in Zygostates in which the outer lateral staminodes are absent, but the whole of the inner ones are fully and largely developed. The structure of Pterostylis enters into the same category, although in some respects very different.

No. 8 may be regarded as the expression of Maxillaria, with all the staminal apparatus gone except the usual outer stamen and the corresponding inner staminode in the form of a tumour on the lip.

No. 9, with every part wanting except the outer central stamen, shows what the structure is of many Dendrobes, and Sarcopods.

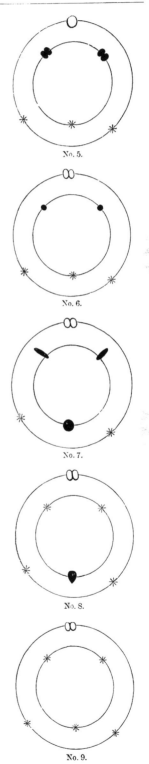

No. 5.

No. 6.

No. 7.

No. 8.

No. 9.

But although these differences exist, and notwithstanding their seeming importance, we own our inability to discover their true value. It does not appear that they can even be employed for the limitation of genera; for *Odontoglossum læve* can hardly be said to possess a trace of the great staminodes of both series which are generally characteristic of genera. This question is, however, only now opened, and it may happen that further observations from this point of view may show a means of employing staminodial distinctions at present unsuspected.

GLEANINGS AND ORIGINAL MEMORANDA.

616. CALCEOLARIA CHELIDONIOIDES. *Humboldt, Bonpland, and Kunth.* A very pretty half-hardy annual, native of Peru. Belongs to Linariads. Flowers yellow. Introduced by Isaac Anderson, Esq., of Edinburgh. (Fig. 300.)

A decumbent, branching, entangled, viscid, hairy, brown-stemmed annual. Leaves pinnated, with pedicellate lanceolate incised divisions, the uppermost ternate, the lowest of three or four pairs with an odd leaflet very much larger than the others. Flowers in pairs, in the axils of every one of the upper leaves, on slender stiff stalks covered closely with spreading brown glandular hairs, as also is the calyx, the lobes of which are incised. Corolla hairy externally, small, but a brilliant pure yellow; its upper lip hardly so long as the calyx, the lower lip obovate and nearly sessile. Anthers with the connective in the form of two horizontal arms, forming a right line at right angles to the filament; the back arm concealed beneath the upper lip of the corolla and antheriferous; the anterior arm longer, hornlike, clear yellow, prominent, and sterile. Seeds very small, smooth, cinnamon-coloured, oblong, strongly-ribbed. With the same kind of treatment as the small Blue Lobelias, it flowers all the summer and autumn, if planted in rather a moist situation. It is very pretty, and a most abundant flowerer, well suited for planting in the American border.—*Journ. of Hort. Soc.,* vol. vii.

300

617. BEGONIA HERNANDIÆFOLIA. *Hooker.* A very fine hothouse herbaceous plant. Native of Veragua. Flowers deep rose. Introduced at Kew.

Received at the Royal Gardens from seeds sent by Mr. Seemann. It is a most lovely species, with singularly shaped, very thick, concave and peltate leaves, deep blood-colour beneath, and the copious petioles, peduncles, and flowers of a full rose-red. It flowers readily in the stove during the summer months. Stemless. From the top of the root spring numerous bright red terete petioles, stipuled at the base, two to four inches, or rather more, long, which are inserted underneath, and at nearly an inch distance from the base of the very thick, between fleshy and coriaceous, subrotundo-ovate, acuminated, rather oblique, concave leaves, indistinctly glanduloso-serrated at the margin, quite glabrous, dark green above, with a pale spot at the insertion of the petiole, from which a few indistinct nerves radiate, deep blood-red beneath, with the nerves slightly prominent. Scapes radical, longer than the petiole, about as thick and of the same colour as it, bearing a dichotomous corymb of drooping, deep rose-red flowers; at the setting-on of the branches a pair of opposite small stipules are present.   Each fork generally bears one male and one female flower.   Male flower of four spreading sepals, two (opposite) larger and orbicular, the two smaller oblong-spathulate.   Female flower of three sepals, two large, and a small oblong-spathulate one. The fruit (nearly mature) is bright red, triangular, obovate, with a narrow rounded wing at two of the angles, and a much broader rounded one at the third angle.—*Bot. Mag.*, t. 4676.

618. CYMBIDIUM GIBSONI. *Paxton.*
A terrestrial Orchid, from the Khasiya hills.   Flowers sweet, greenish, with brown spots.   Introduced by his Grace the Duke of Devonshire.   (Fig. 301.)

C. caule fusiformi articulato nudo, foliis lanceolatis acutis, spicis lateralibus strictis paucifloris, scapo squamis vaginantibus parum longiore, sepalis linearibus obtusis apice latioribus, petalis erectis obtusis sepalo dorsali paulò brevioribus, labello ovato medio contracto apice recurvo obtuso, lamellis 2 arcuatis clavatis continuis.

This little-known plant flowered at Chatsworth last March. It seems naturally allied to *C. ensifolium*, and *lancifolium*, and is readily recognised by its fusiform jointed naked stem, and lateral inflorescence, unusual circumstances among Cymbids.   The species is of little importance as an ornamental plant.

619. CENTROSOLENIA BRACTESCENS. *Hooker.* (*aliàs* Nautilocalyx hastatus *Hort.*) A hothouse herbaceous plant belonging to the Order of Gesnerads. Native of New Grenada or Venezuela (?). Flowers white.   Introduced by Mr. Linden.

We adopt the opinion of Mr. Bentham in considering the genus to which the plant belongs not distinct from his Centrosolenia.   From every known species, the present is abundantly distinguished by the large size of the leaves, and, in proportion, the still larger size and peculiar form of the external bracteas, which enclose the axillary clusters of leaves.   It is a stove-plant, a free flowerer, and its blossoms continue to appear through the entire summer months.   Stem stout, herbaceous, erect, simple, two feet high, the upper part clothed, as is most of the younger portion of the plant, with deciduous silky down. Leaves opposite, very large (almost a foot long), nearly equal, ovate, acuminate, coarsely serrated, penninerved, beneath reticulated and the nerves prominent, below tapering very much ; the base of the two opposite leaves unite and surround the stem, or, in other words, the leaves are decurrent upon the petiole so as to form a very broad wing to the extremely thickened rachis.   In the axils of the leaves there appears on a short peduncle a very large, vertical, nearly orbicular, concave, sharply almost cuspidately acuminated, purple-green reticulated bractea, two inches across, at first closed like the two valves of such a shell as a Pecten or Venus, then partially expanded

301

for the emission of the several flowers, within which they expand in succession, and are themselves bracteated with ovate or lanceolate acuminated and serrated bracteoles. Each flower, when fully open, is nearly as long as the external bracteas, and shortly pedicellate. Calyx a little shorter than the tube of the corolla, white below, red-purple above, and reticulated with white, deeply cut into five segments, of which four are lanceolate, serrated, finely acuminated, the fifth free to the very base, and bent down, as it were, below, by the prolongation of the spur, and this is subulate, very narrow. Corolla large, white, the tube dilated upwards, below on one side extended into a short, blunt spur; the limb spreading, of five nearly equal, entire, rounded segments or lobes. Stamens four, perfect, included within the tube of the corolla; filaments subulate, didynamous, curved over the pistil. Anther subglobose. Ovary ovate, slightly pubescent, with a large fleshy hypogynous gland on one side. Style thickened, a little curved. Stigma slightly dilated.—*Bot. Mag.*, t. 4675.

620. LOPEZIA MACROPHYLLA. *Planchon.* (*aliàs* Jehlia fuchsioides *Hort.*) A showy half-hardy perennial. Flowers deep rose. Native of Guatemala. Belongs to Onagrads. (Fig. 302.)

This is a soft smooth pale green shrub, with a fleshy tuberous root, like some Fuchsias. The leaves are stalked, almost wholly smooth, oblong-lanceolate, acuminate, narrowed to the base, where they terminate abruptly in a rounded manner, strongly serrated, and furnished with deep lateral diverging veins, which give the leaves the appearance of a Hornbeam tree's; at the base they are furnished with a pair of red pyramidal short glands. The flowers stand on long slender stalks, singly in the axils of leaves, are as large as those of a *Fuchsia globosa*, and of a very deep rose-colour, which pervades every part except the anther, which is blue. At first sight this plant would not be taken for a Lopezia, the sepals being altogether petaloid, and the glandular knee peculiar to two of the petals of the genus seeming to be absent. But it will be found upon a careful examination that the knees are really present, only they stand very low down on the petals, so as to be concealed by the other parts. The name here employed, but with some doubt, is that under which M. Planchon has given it in the *Flore des Serres;* but it seems impossible that it can be the plant which Mr. Bentham first described as *Lopezia macrophylla*, in the *Plantæ Hartwegianæ*, a shrub with downy leaves and terminal panicles of flowers. That species I suspect exists in the

302

Society's Garden, from Mr. Skinner, but, not having flowered, cannot at present be identified. Till materials accumulate for the satisfactory settlement of this question the name employed by M. Planchon had better stand unchanged. A greenhouse soft-wooded shrub, growing freely in a mixture of sandy loam and leaf-mould, and requiring the same treatment as a Fuchsia. It is increased by cuttings put in sand under a bell-glass, and flowers during winter and spring. It is likely to be valuable as a winter flowering plant, notwithstanding that it is coarse in foliage and habit. —*Journ. of Hort. Soc.*, vol. vii.

621. MYRICA CALIFORNICA. *Chamisso and Schlechtendahl.* A handsome hardy evergreen shrub.

Native of California.   Belongs to Galeworts.   Berries bluish grey.   Introduced by the Horticultural Society.

Said to be wild in woods near Monterey, growing twelve feet high. This was originally gathered by Menzies on the north-west coast of America. Douglas found it at Puget Sound. It forms an evergreen bush, with dense, narrowly lanceolate, slightly serrated leaves, covered, especially on the under side, with transparent, glossy, saucer-shaped, sunken scales, of microscopical dimensions, consisting of a layer of wedge-shaped cells, placed obliquely round a common centre. The flowers are green and inconspicuous, in short axillary spikes, which eventually bear from one to three small globular fruits, whose surface is closely studded with fleshy, oblong, obtuse grains of a dull red colour, and astringent flavour. It is a hardy evergreen, growing freely in any good garden soil, increased by seeds or by layers, in the usual way. It flowers in July, and produces in September an abundance of its little granular fruits. In gardens it is an acquisition, being a hardy shrub, with fragrant leaves, and well suited for rock-work or for the front of a shrubbery. —*Journ. of Hort. Soc.*, vol. vii.

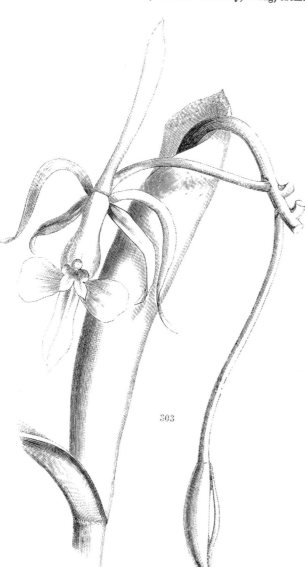

303

622.   EPIDENDRUM LEUCOCHILUM. *Klotzsch.* (*aliàs* E. flavidum *Lindl.*) A handsome epiphyte from New Grenada. Flowers large, yellowish, with an ivory white lip. Exists in the German Gardens. (Fig. 303.)

This is a fine caulescent fleshy-leaved species, with the habit of *E. umbellatum*, and such flowers as those of *E. nocturnum.* The stem is about two feet high. Leaves coriaceous, distichous, recurved, emarginate. Raceme many-flowered, drooping, issuing from a long green compressed spathe. Flowers three inches in diameter, upon stalks rather shorter than themselves. Petals and sepals green, in Germany according to Dr. Klotzsch, yellowish in its native country according to Mr. Linden. Although as fine a species as *E. ciliatum*, this does not seem to have yet reached our English gardens. When the *Orchidaceæ Lindenianæ* were published, I only knew the plant by the specific character and description given of it in *Allgemeine Gartenzeitung*; and I then supposed it to be different from what an excellent figure in the *Icones Berolinenses* shows it to be. Under this misapprehension, when I found it among Mr. Linden's Orchids (No. 2213), I supposed it to be new, and called it *E. flavidum*, an error which is now corrected.

623.   ASTRAGALUS PONTICUS. *Pallas.* A hardy herbaceous plant of the Leguminous Order. Flowers yellow. Native of the West of Asia. Introduced by H. C. Calvert, Esq., of Erzeroom.

A decumbent perennial of a bright lively green colour. Stems about two feet long, slightly downy. Leaves almost smooth, of the texture of the Garden Pea, about a foot long, composed of seventeen or eighteen pairs of ovate-oblong, obtuse, or emarginate leaflets. The flowers are bright yellow, in nearly sessile ovate heads, with short calyx tube, much less hairy than in the allied species. The cultivators of hardy herbaceous plants will understand what this is when it is compared with *Astragalus alopecuroides*, which it is a good deal like. It is half-shrubby, growing freely in peat-soil, and flowering in August. Though not showy, its fine foliage renders it well

adapted for shrubberies coarse rockwork, and flower borders, devoted to the cultivation of the rougher kind of perennials.—*Journ. of Hort. Soc.*, vol. vii.

624. BOMARIA ACUTIFOLIA. *Herbert.* (*aliàs* Alstrœmeria acutifolia *Link and Otto.*) A half-hardy twining herbaceous plant. Native of Mexico. Flowers dull red. Belongs to Amaryllids. (Fig. 304.)

Stem, according to M.M. Link and Otto, attaining a height of five or six feet, somewhat twining, rounded, glabrous. Leaves remote, lanceolate, much and narrowly acuminated, striated, dark green and glabrous above, pale and downy, especially on the nerves beneath, inserted on a short, twisted petiole. Umbel terminal, of many flowers, surrounded at the base by an involucre of leaves, resembling those of the stem, but much smaller. Peduncles rounded, flexuose, downy. Corolla sub-campanulate ; the petals erect, and but slightly patent at the extremity, all of them nearly equal in height, the three outer ones oblong, of a deep but not very bright red,

acute ; the three inner more delicate in texture, broadly spathulate, orange-coloured ; all of them with a deep red spot at the tip. Stamens shorter than the corolla. Filaments pale reddish-purple. Anthers oblong, bluish purple. Germen inferior, turbinate, triangular, downy ; style straight, filiform, greenish white, thickened at the base, at the extremity terminated by a trifid stigma. Capsule remarkably depressed, turbinate, opening at the top by three valves, each of which bears a septum, and each septum has two seeds attached on either side of it, of a bright scarlet colour.

For this beautiful species of Alstrœmeria our gardens are indebted to Mr. Otto of Berlin, who transmitted plants to the Botanic Gardens both of Edinburgh and Liverpool ; and from specimens that have flowered in both those establishments, the present figure and description have been made.—*Bot. Mag.*

It inhabits Mexico, where it was discovered by M. Deppe, who is now most successfully exploring that interesting region as a Botanist. In our stoves it has flowered in the months of August and September.—*Bot. Mag. t. 3050.*

The greater part of these beautiful plants are natives of elevated situations and dislike a high temperature. They will be found to thrive best out of doors in this country in summer time, and will endure the winter if planted pretty deep in light soil and covered over with leaves in the cold season, especially if any sloped heading be laid on to throw off the wet. Even *acutifolia*, which in the greenhouse keeps its leaves through the winter, will succeed with that treatment.—*Herbert. Amaryllid.* p. 120.

625. TACSONIA SANGUINEA. *De Candolle.* (*aliàs* Passiflora sanguinea *Smith; aliàs* P. diversifolia *of Nurseries; aliàs* P. quadriglandulosa *Meyer; aliàs* Tacsonia quadriglandulosa, T. quadridentata (?) *et* T. pubescens (?) *De Candolle,* according to Hooker.) A very fine hothouse creeper, with large scarlet flowers. Native of Trinidad. Blossoms in July. Introduced by Messrs. Low and Co.

Unquestionably the *Passiflora sanguinea* of Sir J. E. Smith, in *Rees's Cyclopædia,* and only by that description known to De Candolle, who was induced to refer the species, in its present genus, to the section *Eutacsonia ;* and thus, apart from three supposed West Indian species, *T. quadriglandulosa, T. quadridentata,* and *T. pubescens,* placed in the section " Distephanæ dubiæ." These three, though very briefly characterised by De Candolle, one from Guiana (whence we have also received this species), and the two others from the " West Indies," derived from the Banksian Herbarium, and very probably from Trinidad, appear to us to be referable to one and the same plant. The very variable nature of the leaves on the same or on different individuals will easily account for their being supposed distinct. Mr. Low observes that the species is a free flowerer, and will evidently make a first-rate plant for a conservatory, as it does not seem to require much heat, and is easy of cultivation. A climber, with terete branches, and leaves which are extremely variable on the same or on different plants, sometimes ovate or oblongo-ovate, acute, simple ; sometimes cordate and deeply three-lobed, with the lobes ovate, acute; the margins everywhere remarkable for being more or less sinuous, and cut into large but unequal teeth, penninerved, the underside strongly reticulated with prominent nerves, sometimes downy and pale green, whereas the upper side is generally glabrous and dark green. Petioles about half an inch long, glandular at the base, and there are sometimes glands in the sinuosities of the leaves. Peduncle solitary, single-flowered, longer than the petiole, furnished below the apex with a small three-leaved downy involucre : the leaflets from a broad base, linear-subulate, serrated, erect, each having one or two large orbicular glands on either side at the base, and a gland within the axil. Flower large : sepals five, oblong-linear, acuminate, spreading, having a long soft subulate awn a little below the apex ; externally the sepals are greenish rose-colour, within uniform rose-red : they all unite below so as to form a five-furrowed, rather short, greenish tube, very obtuse at the base. Petals five, as long, and of the same shape, as the sepals, equally spreading, and deep rose-red on both sides. Crown or nectary double, short : inner consisting of a white membrane, with many subulate, erect, red rays ; outer of a circular row of numerous erect filaments, white, banded and tipped with red : some lesser filaments, and very short, are found between the outer and inner corona. Column three or four times as long as the crown, greenish, spotted with red, as are the short recurved filaments. Anthers green. Ovary oval. Styles clavate, deep red ; stigmas green.—*Bot. Mag., t. 4674.*

626. VANDA LONGIFOLIA. *Lindley.* An unimportant hothouse epiphyte, with yellow flowers. Native of the East Indies. Introduced by the Court of Directors of the East India Company.

This is a very fine-looking plant when not in flower, with dark green distichous leathery wavy leaves, as much as a foot and a half long and two inches wide, obliquely rounded at the end. Its habit is almost that of *Angræcum eburneum.* Very thick greyish-green roots protrude from its stem, and have a tendency to branch wherever the first point is injured. But the flowers are insignificant, very much like those of *Vanda multiflora* in form and colour, except that they are paler ; they, however, have a pleasant perfume. These flowers appear in a corymb at the end of a short stiff ascending peduncle not one quarter the length of the leaves ; they are very fleshy, and are banded with red upon a dull yellow ground ; the lip is white. Inside the pouch of the lip are numerous yellowish hairs, concealing an erect fleshy plate, which partially divides the hollow of the lip into two halves. It is not worth cultivating for the flowers, but the foliage is handsome, and serves to set off other Orchids.—*Journ. of Hort. Soc.,* vol. vii.

627. CEANOTHUS VERRUCOSUS. *Nuttall.* A very valuable hardy evergreen shrub. Native of California. Flowers light blue, in June. Belongs to Rhamnads. Introduced by the Horticultural Society as " a shrub eight feet high, growing on the Santa Cruz mountains."

This proves to be a hardy evergreen of the best kind. It forms already a large bush, and will probably become a tree with long stiff rod-like downy branches, covered in winter with multitudes of large oblong or roundish brown buds. The leaves are opposite, roundish oblong, either slightly notched or entire at the end, scarcely an inch long at the largest, flat, deep green, shining, with grey hairy pits distributed over all the under surface. Occasionally, when the plant is young, they are coarsely toothed, as is represented in the *Botanical Magazine ;* but that is an exceptional state.

At the base of each leaf is a pair of stipules, which gradually lose their thin extremities and change into soft fleshy conical prickles. The flowers are very pale blue, produced in great abundance in dense corymbs at the end of very short stiff lateral branches. This shrub is among the most easy of plants to grow, and seems indifferent to climate or soil. It is increased by cuttings of the half-ripened wood, placed in sand under a hand-glass in a north aspect about the end of August. It is, however, best propagated by layering in the autumn. It flowers in June. It may be added that with the single exception of *C. cuneatus*, a white-flowered species of little beauty, all the Californian Ceanothuses prove to be hardy near London. It is only requisite that they should not be placed in soil which keeps them growing till late in the year, but that their wood should be well ripened. In the *Botanical Magazine* Sir William Hooker, in speaking of *C. rigidus*, observes that—" The North-west American Ceanothuses are particularly deserving of cultivation in the open ground ; but it may require a Devonshire climate to bring them to the state in which they are at Bishopstowe, as just announced to me in a letter, dated 27th May, 1852, of the Bishop of Exeter :—' The

305

Ceanothus divaricatus is now in its highest beauty : the largest plant is eighteen feet high, eighteen feet wide, twelve feet deep (*i. e.* from back to front), covered with thousands of the beautiful thyrsoid flowers, so that the leaves are hardly visible. *C. rigidus* blossomed about six weeks ago ; *C. dentatus* is now in full flower ; *C. papillosus* is just coming into flower ; *C. azureus* will not blossom before August.' "—*Journ. of Hort. Soc.*, vol. vii.

628. Eugenia? apiculata. *De Candolle.* An evergreen half-hardy shrub, from Chili. Flowers white. Fruit deep purple. Belongs to Myrtleblooms (*Myrtaceæ*). Introduced by Messrs. Veitch & Co. (Fig. 305.)

This is a plant with much the appearance of the common Myrtle. The branches are clothed with rusty hairs. The leaves are roundish ovate, sharp-pointed, downy on the under side when young, but quite smooth and deep green when old. The flowers, which are solitary and axillary, consist of four white concave petals uneven at the edge, outside which stand four leafy round sepals. The fruit is a spherical purple berry, the size of that of the common Myrtle, with a pair of extremely minute bracts at the base, and crowned by large green sepals. It contains from one to two fleshy seeds, enclosed in a thin tough skin, with a long thick cylindrical radicle folded down upon the outside of a pair of plano-convex cotyledons, which are either flat or more or less folded together. It is an exceedingly pretty shrub for the milder parts of England. But to what genus does it belong ? The seeds, which are very like those of Vicia Faba on a small scale, correspond with no generic character yet published. According to De Candolle the seeds of Myrtus have a bony shell and a curved embryo with long semicylindrical cotyledons ; to Eugenia is assigned an embryo with consolidated cotyledons and a very indistinct straight radicle ; Jambosa has the radicle enclosed between the cotyledons. With none of these will the plant before us agree ; yet we cannot believe that it is a new genus. It is better to suppose, what can hardly be doubted, that the generic characters of the fleshy-fruited Myrtaceæ are greatly in need of amendment.

629. Tillandsia stricta. *Botanical Magazine.* A hothouse epiphyte with blue flowers. Native of Brazil. Belongs to Bromeliads.

This is a small Pine-Apple-like plant, about six inches high when in flower. The leaves are very narrow, channelled, mealy, stiff, terminating in a long drawn-out point, and curved backwards till their ends are below the base of the plant. The flowering stems are shorter than the leaves, curved downwards, clothed with small green leaves resembling those below them. The flowers are collected into oblong cones, formed of shining, naked, roundish ovate, convex, imbricated bracts, the lower of which have a leafy point.   Two varieties were observed, one with bright rose-coloured bracts and blue flowers, the other with greenish bracts and white flowers.   Among the less important inhabitants of the stove this may be regarded as a useful little plant, growing best in a warm moist air, attached to a block of wood, where it flowers in August.—*Journ. of Hort. Soc.*, vol. vi.

630. ECHEVERIA QUITENSIS.   *Lindley.*   (*aliàs* Sedum quitense *Humboldt and Kunth.*)   A very pretty half-hardy succulent plant.   Native of Peru.   Flowers deep red.   Belongs to the Order of Houseleeks.   Introduced by Isaac Anderson, Esq. of Edinburgh.

A bright green smooth succulent plant, forming stiff erect stems about six inches high, clothed by imbricated spathulate leaves, with an almost circular base attached to the stem only by one bundle of fibro-vascular tissue.   The flowers are in stiff close erect racemes, shorter than the lower bracts, which resemble in form the leaves, but taper less to the base.   Sepals five, longer than the pedicel, equal, linear, acuminate, rather shorter than the corolla, which forms a scarlet five-sided pyramid, opening very slightly at the end into five acuminate lobes.   Of the ten stamens, five stand in furrows of the petals, and five are distinct.   This is evidently an Echeveria, as De Candolle surmised, and not a Sedum.   During the summer it does very well on rockwork out of doors, but it is probable that it should be treated as a green-house shrubby succulent plant, requiring the same kind of soil and treatment as Echeverias.   It is easily increased by cuttings, and seeds, which it ripens abundantly.   When grown out of doors, though pretty, it is not a very striking plant.   It flowers in August.   How it will look in a greenhouse is not ascertained as yet.—*Journ. of Hort. Soc.*, vol. vii.

631. VINCETOXICUM JAPONICUM.   *Morren and Decaisne.*   (*aliàs* Cynanchum flavescens *Siebold.*) A hardy herbaceous plant from Japan.   Flowers pale yellow.   Belongs to Dogbanes.   (Fig. 306.)

A herbaceous plant, with a slight tendency to climb.   The whole surface soft with down.   Leaves roundish, oblong, mucronate, nearly sessile.   Flowers few, pale greenish-yellow, in nearly sessile cymes, with slender pubescent flower-stalks.   A perennial, supposed to be hardy or half-hardy, growing best in the peat border, and increased by division of the roots when in a dormant state.   It is, however, of no kind of horticultural interest.   It flowers in July and August. —*Journ. of Hort. Soc.*, vol. vii.   We figure this just for the sake of showing what sort of things are sometimes sent to this country as new and valuable *Garden* plants.

306

PLATE 1

L.Constans del.& zinc.

Printed by C.F.Cheffins, London.

[Plate 103.]

THE GOLDEN-FLOWERED DIELYTRA.

(DIELYTRA CHRYSANTHA.)

———◆———

A handsome hardy Herbaceous Plant, from California, *belonging to the Order of* Fumeworts.

THE GOLDEN-FLOWERED DIELYTRA. Stem tall, leafy, branching. Leaves twice or thrice pinnate, with linear acute smooth segments. Panicle long. Bracts and calyxes broad-ovate, blunt. Petals spathulate, the outer scarcely gibbous at the base; the inner with a broad wing along almost the whole length of the back. Stigma very broad, truncate.

DIELYTRA *CHRYSANTHA ;* caule elato folioso ramoso, foliis 2—3-pinnatim sectis segmentis linearibus acutis glabris, paniculâ elongatâ, bracteis calycibusque latè ovatis obtusis, petalis spathulatis exterioribus basi vix gibbosis, interioribus dorso ferè per totam suam longitudinem lato-alatis, stigmate latissimo truncato.—*Hooker and Arnott.*

Dielytra chrysantha : *Hooker and Arnott, Botany of Beechey's voyage,* p. 320, t. 73.

This very handsome hardy perennial was originally discovered in California by Douglas, from whose specimens it was published in the work above quoted. More recently it has been found in the same country by Mr. W. Lobb, from whose seeds Messrs. Veitch succeeded in raising it. It flowered in the Exeter nursery for the first time last September.

It forms a handsome tuft of firm very glaucous foliage, sometimes much more finely cut than in our figure, and in general texture and colour resembling Garden Rue. Among the leaves rise stiff branching panicles of rich golden-yellow blossoms.

Although very inferior to *Dielytra spectabilis,* this has a beauty of its own, which will render it a

Y

favourite for autumn decoration. The contrast between the gray dull leaves and gay glittering flowers is particularly agreeable.

We are not aware that this demands any particular care. Like other Californian plants it likes a roasting summer, and therefore should have the warmest and driest berth which the garden can afford.

PLATE 104.

L. Constans del. & zinc.   Printed by C.F. Cheffins, London.

[PLATE 104.]

THE BELL-FLOWERED SPATHODEA.

(SPATHODEA CAMPANULATA.)

◆

A magnificent Hothouse Shrub, from TROPICAL AFRICA, *belonging to the Natural Order of* BIGNONIADS.

Specific Character.

THE BELL-FLOWERED SPATHODEA. A tree, apparently smooth. Leaves alternate, unequally pinnate; the leaflets of four pairs, lanceolate, quite entire. Raceme terminal, somewhat branched. Calyx velvety in longitudinal lines, curved at the point. Corolla campanulate, smooth, with a nearly equal limb.

SPATHODEA *CAMPANULATA*; arborea glabra (!), foliis alternis impari-pinnatis, foliolis 4-jugis lanceolatis integerrimis, racemo terminali subramoso, calyce longitudinaliter subvelutino nervoso apice arcuato, corollâ campanulatâ glabrâ limbo subæquali.—*De Candolle.*

Spathodea campanulata : *Palisot de Beauvois, Flore d'Oware et de Benin,* I. 47, t. 27 ; *De Candolle, Prodrom.* 9. 208 ; *Bentham, in Hooker's Niger Flora,* p. 461 ; *aliàs* Spathodea tulipifera : *G. Don ; aliàs* Bignonia tulipifera : *Schumacher and Thonning, Beskryving,* p. 273.

THIS gorgeous plant produced its flowers at Chatsworth last August, when the accompanying figure was made. It had previously blossomed in June, at which time the flowers were still finer, and the colours more distinct and rich. It has a fine Ash-like habit, producing great opposite pinnate leaves, with broad leaflets, from among which come the glorious racemes of Tulip-like tough leathery fiery-orange flowers, six or seven together; they are quite as handsome as the wild specimens before us from the Niger, where it was found on Stirling Hill by poor Ansell.

Palisot de Beauvois says it is a middle-sized tree, with wood smelling strongly of garlic when broken. He only found one specimen three leagues north of Chama.

In the *Niger Flora* Mr. Bentham speaks of the plant thus :—

"Although the descriptions differ in several points, there is every reason to conclude that Beauvois' and Thonning's plants belong to one species. Beauvois' characters are generally drawn up from mere fragments, his drawings made on the spot of this and other plants having been destroyed by fire at St. Domingo, and he is very likely to have committed the mistake of describing the leaves as alternate instead of opposite. The corollas in Ansell's specimens are fully as large as that figured by Beauvois ; those which are well dried, are even larger ; Thonning says they are as large as the largest tulips. The leaflets in Ansell's plant are rather broader than in Beauvois' ; they are covered on the underside with a minute tomentum, which is scarcely perceptible in the older leaves; they are also marked on the same side with innumerable small black dots, only visible under a lens. Thonning's detailed description is very accurate."

We believe the introduction of this plant to our gardens is owing to Mr. Whitfield, well known as an indefatigable collector of objects of natural history in Tropical Western Africa.

PLATE 105.

a

b

L.Constans del.& zinc.

Printed by C.E.Cheffins, London.

[PLATE 105.]

THE HAYTIAN LÆLIOPS.

(LÆLIOPSIS DOMINGENSIS.)

———◆———

A handsome Hothouse Epiphyte, from St. Domingo, *belonging to the Natural Order of* Orchids.

Generic and Specific Character.

LÆLIOPS. A Cattleya in all respects, except that the flowers are membranous, and the veins of the lip bearded.

THE HAYTIAN LÆLIOPS. Pseudobulbs 2-leaved. Leaves oblong, coriaceous, obtuse. Scape slender, naked, with about 8 flowers at the end. Lip 2-lobed, with its divisions wavy, denticulate, recurved. Central veins bearded.

LÆLIOPSIS. Omnino Cattleya, nisi quod flores membranacei necnon venæ labelli barbatæ.

LÆLIOPSIS *DOMINGENSIS;* pseudobulbis 2-phyllis, foliis oblongis coriaceis obtusis, scapo gracili nudo apice sub 8-floro, labelli 2-lobi laciniis denticulatis undulatis recurvis venis centralibus barbatis.

Cattleya domingensis : *Lindl. Gen. & Sp. Orch.*, p. 118 ; Broughtonia lilacina : *Henfrey, in Gardener's Magazine of Botany,* Vol. III., p. 201, *with a figure.*

What is the genus of this beautiful plant ? Lælia ? no ; because it has only four pollen-masses—Broughtonia ? no ; for although its flower is deeply cuniculate, yet it has not a long external adnate spur and decurrent sepals—Epidendrum ? no ; for it wants the unguiculate lip more or less united to the column—Cattleya ? still no ; although we once thought it one ; for the flowers are membranous, the veins of the lip bearded, and the habit quite different.

We see no means of providing a fixed station for this and a few allied plants, except by giving them a genus to themselves, the essential features of which shall consist in what has been above proposed. There is no doubt that Cattleya, Epidendrum, and Broughtonia, are so very nearly related that on mere technical grounds they might be all placed in the same genus : but their habits are very different, and the mind is unable to reconcile itself to their union.   As to BROUGHTONIA, if we disregard its cucullate lip and manifest external adnate spur, there is little to divide it from Epidendrum, the majority of whose species have a cuniculate ovary, and in the case of *E. vesicatum*, even a spur partially visible ;—or from CATTLEYA, except the tough coriaceous quality of the lip in that genus, and the adhesion of the sepals of BROUGHTONIA to the face of its external spur. Upon grounds of the same nature as those which separate these genera must LÆLIOPSIS be sustained, when the mutual differences among the four genera may be tabulated thus :—

Labellum calcaratum, sepalis calcari adnatis.   BROUGHTONIA.
Labellum ecalcaratum, cuniculatum tantum.
    unguiculatum ; ungue sæpius columnæ adnato.   EPIDENDRUM.
    sessile, convolutum.
        coriaceum imberbe.   CATTLEYA.
        membranaceum barbatum.   LÆLIOPSIS.

    Læliopsis thus defined will receive, in addition to the species now published, *Lælia Lindenii*, *Broughtonia chinensis*, and *Epidendrum cubense*.

    Læliopsis domingensis was first found on trees in St. Domingo, by Mr. Mackenzie; then Jaeger gathered it off branches of the Logwood tree in woods near Miragoane, where he saw it in flower in April.   It has lately been introduced to our gardens, and exhibited by Messrs. Henderson, of Pine Apple Place Nursery, and Mr. Rucker.   We owe our opportunity of making a drawing to W. F. G. Farmer, Esq., of Nonsuch Park, who sent it us in the course of last summer.

    It is an extremely pretty species, because of its gay lilac flowers a little veined with yellow in the middle of the lip.   Like other St. Domingo plants it demands all the heat of the stove while growing; but it appears to be naturally dried up after the growth is made, if we are to judge from our wild specimens.

GLEANINGS AND ORIGINAL MEMORANDA.

632. SALVIA HIANS. *Bentham ; var.* plectranthifolia. A hardy herbaceous plant belonging to Labiates. Native of the Himalayas. Flowers violet and white. Introduced by Major Madden. (Fig. 307.)

We received a specimen of this in October last from Mr. Moore, the superintendent of the Glasnevin Botanic Garden, with the following note :—

"It is pretty, and quite hardy. Major Madden collected the seeds, from which I raised the plants, near Simlah. He told me lately that Mr. Bentham admits it to be quite a new species to him. In its native habitat, I understand, *Salvia plectranthifolia* is a very showy species, and conspicuous among the hill plants of that country."

We nevertheless confess our inability to find a good specific difference from the *S. hians*, figured in the *Botanical Register*, 1841, t. 39. The specimen sent us had rather smaller and less hairy flowers, and the leaves were rather more obtuse at the base ; the flowers too were rather more violet, but we saw nothing more peculiar. At all events it is a very pretty hardy herbaceous plant, gay with violet flowers, having a pure white centre to the lip.

307

633. ROSA FORTU-NIANA. *Lindley.* Fortune's Double Yellow, or Wang-jang-ve Rose.

If it is desirable to give a botanical specific name to a hybrid plant at all, it can only be done, with any kind of propriety, when we are acquainted with the double origin of the plant in question, viz. both parents. Of the pedigree of the Rose here figured we know nothing, save that it comes from China; and, as Dr. Lindley has observed, it is fruitless to inquire. As an ornamental rose for the garden, we should have thought there could have been but one opinion among those who have seen the flowering plant (the delicacy of the petals cannot be imitated by art), and that is entirely in its favour. But it has been spoken of unfavourably by some; and this has been accounted for by Messrs. Standish and Noble, to whom we are indebted for the specimens here published, and whose remarks, together with those of Mr. Fortune, who introduced the plant from China to our gardens, shall occupy the remainder of our space. "Seldom," write Messrs. Standish and Noble, in June of the present year, "has a really beautiful flower remained so long comparatively unknown as this. Few persons have seen a blossom; and those who have not, believe it to be worthless. In fact there exists a deeply rooted prejudice against the plant, caused, no doubt, by the very unfavourable report circulated when it bloomed the first time in this country. Yet nothing can be more beautiful as a flower, nor can anything exceed it in delicacy of tint. Imagine a gamboge-yellow ground, over which is thrown a tint of crimson lake, and you obtain an idea of its colour. The centre petals have generally a predominance of lake, and the outer ones are more strongly marked; but there is a beautiful clearness about them, which can only be appreciated by examining a flower. Apart from the prejudice which exists against the plant, many persons have spoken derogatively of it, from having failed to cultivate it successfully; their plants producing but few flowers, and those indifferent both in size and colour. This has arisen from an improper mode of treatment. If pruned in the manner usually adopted for ordinary standard roses, no flowers will be obtained, as they are produced from the wood of the preceding year, in the same manner as those of the Persian yellow and Banksian roses. Therefore, whether grown as a standard or trained to a wall, the shoots should only be thinned,—to shorten them is to destroy the flowers. We have at the time of writing this (June 28) some standards, from three to four feet through the heads, covered with blossoms; and more beautiful objects can scarcely be imagined. We wish all who are prejudiced against the plant could see them. Again, it has been said to be tender; but we have never seen it injured in the least, even during the most severe weather. It is one of the most rapid-growing roses, and well adapted for a wall or pillar." Mr. Fortune tells us, "the rose you inquire about is well known to me, and was discovered in the garden of a rich Mandarin at Ningpo. It completely covered an old wall in the garden, and was in full bloom at the time of my visit: masses of glowing yellowish and salmon-coloured flowers hung down in the greatest profusion, and produced a most striking effect. It is called by the Chinese the Wang-jang-ve, or yellow rose. They vary, however, a good deal in colour; a circumstance which, in my opinion, adds not a little to the beauty and character of the plant. I fancy it is quite distinct from any other known variety, and certainly different from any China kind. It is admirably adapted for covering walls: and if planted in rich soil, and allowed to grow to its full size, nothing can produce a finer effect in our gardens. It was sent home to the Horticultural Society in 1845, and noticed by me in the Journal of the Society, vol. i. p. 218, and again in my *Journey to the Tea Countries*, p. 318. No doubt the Wang-jang-ve, now that it has been properly treated by Messrs. Standish and Noble, will soon take its place as a favourite amongst our climbing roses."—*Bot. Mag.*, t. 4679.

634. COMACLINIUM AURANTIACUM. *Scheidweiler.* (*aliàs* Tithonia splendens *Gardens.*) A most beautiful half-hardy perennial, with scarlet flower-heads. Native of Guatemala (?). Belongs to Composites. Introduced by Mr. Van Houtte.

This plant looks like an "African Marigold," with the flowers of a Scarlet Zinnia. M. Planchon says that it was raised from seeds found in the earth belonging to a lot of Orchids from Central America, by Mr. Ortgies, the foreman in Mr. Van Houtte's hothouses. It was planted out under the wall, in front of an Orchid-house, and flowered last autumn. Messrs. Planchon and Scheidweiler are of opinion that it forms an entirely new genus in the tribe of Tagetineæ, which are brought, by their copious oil cysts, close to the Pectideæ, "now lost, one hardly knows why, among Vernoniaceæ." The following is the character given by these gentlemen of their new genus :—

Capitulum multiflorum, heterogamum, floribus radii ligulatis, uniseriatis, fœmineis, disci hermaphroditis (?) tubulosis, centralibus subabortivis. Involucri squamæ circiter 12, subuniseriatæ, à basi liberæ, marginibus tantum leviter imbricatæ, lineari-spathulatæ, infernè longitudinaliter nervosæ, apice dilatato membranaceæ, vittis oleo-resiniferis lineatæ. Receptaculum conicum, floribus avulsis ob paleas in fimbrillas permultas divisas quasi comosum (unde nomen). Corolla flosculorum apice vix dilatata, acutè 5-fida, divisuris crassis, æstivatione subimbricato-valvatis leviterque contortis. Antheræ ecaudatæ : pollen globosum, undique echinulatum. Styli bifidi cruribus ad margines minutè papillosis, sub apice conico breviter annulato-barbatis. Ovaria cylindraceo-clavata, haud manifestè angulata nec alata. Pappus è paleis circiter 12—15, irregulariter biseriatis, infra medium in fimbrillas piliformes, asperas, fissis. Achenia. . . . . Herba Americæ centralis tropicæ, perennis, basi lignosa, 1—3-pedalis, trichotome ramosa, Helianthorum facie. Rami graciles, sulcato-striati. Folia opposita, paribus longiusculè dissitis, in petiolum utrinque setis herbaceis ornatum angustata, lanceolata, acuminata, argutè serrata, supra glaberrima, subtus pilis minutis substrigillosis tactu leviter asperata, cryptis pellucidis punctiformibus conspersa. Capitula terminalia, solitaria, pulchrè aurantiaca, pedunculo longiusculo, apice sensim incrassato, bracteolis 2—3 lineari-lanceolatis involucro admotis.—*Scheidweiler and Planchon.*

635. LILIUM CANADENSE. *L.; var.* occidentale. A fine showy bulbous plant, with narrow

whorled leaves, and revolute orange flowers spotted with crimson. Native of California. Introduced by the Horticultural Society. (Fig. 308.)

"My North-West American specimens of *L. canadense*," writes Sir W. Hooker (*Flor. Bor. Americana*, ii. 181) "have much smaller flowers, of a redder hue, and are more disposed to be revolute; yet I dare not venture upon making them distinct." The plant now figured is, we presume, what he thus referred to. It is remarkable for having long grassy leaves, as many as ten in a whorl, instead of five, which is the usual number in *L. canadense*. The flowers are deep orange, very much like those of a Martagon, with red stains, and numerous rich red-brown blotches. The stamens are not at all united at the base. The plant is very handsome.

636. Fuchsia miniata. *Planchon and Linden.* A very pretty greenhouse shrub, with long tubular rich red flowers, having green-tipped sepals a little longer than the scarlet petals. Native of New Grenada. Introduced by Mr. Linden.

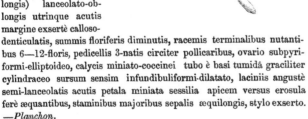

308

F. miniata (§ longifloræ); frutescens, undique (petalis genitalibusque exceptis) puberula (novellis exsiccatione canescentibus), ramis teretibus petiolis denticulisque foliorum rubidis foliis 4- v. 3-natis longiusculè petiolatis (1—2 poll. longis) lanceolato-oblongis utrinque acutis margine exsertè callosodenticulatis, summis floriferis diminutis, racemis terminalibus nutantibus 6—12-floris, pedicellis 3-natis circiter pollicaribus, ovario subpyriformi-elliptoideo, calycis miniato-coccinei tubo è basi tumidâ graciliter cylindraceo sursum sensim infundibuliformi-dilatato, laciniis angustè semi-lanceolatis acutis petala miniata sessilia apicem versus erosula ferè æquantibus, staminibus majoribus sepalis æquilongis, stylo exserto. —*Planchon.*

There is an obvious resemblance between this and *F. venusta*; it has the same verticillate leaves, the same pendent flowers, the same taper-pointed calyx, and the same colour in the petals. The form, however, of the latter is of itself sufficient to distinguish the two plants. There is less, though still abundant, difference between this *Fuchsia miniata* and *F. petiolaris* of Humboldt. In *miniata* the leaves are as often in fours as threes, they are more acuminate, and more evidently downy; the sepals are narrower and less cuspidate, the petals entirely smooth, and obtuse instead of being, as in *F. petiolaris*, acute and furnished with scattered hairs. Slight as these differences may appear, they are not the less to be depended upon, as we learn from a comparison of authentic specimens of the two. To say that *F. miniata* is a worthy rival of *F. venusta*, is to say that it has great merit. Its foliage is not so good, but that fault is made up for by the richness of its flowers. It was found in New Grenada, whence Mr. Linden received it from his collector, Schlim. —*Planchon, in Flore des Serres.*

637. Siphocampylus penduliflorus. *Decaisne.* A very handsome half-climbing stove plant. Native of the Caraccas. Flowers bright deep rose-colour. Belongs to Lobeliads. Introduced by Mr. Linden.

S. penduliflorus; scandens glaberrimus, ramis lignosis supernè angulatis leviter flexuosis epidermide pallide lutescente minutè puncticulatâ vestitis, foliis alternis petioli longiusculi torsione sæpe deflexis ovato-oblongis acutiusculis remotè adpressèque serrulatis crassiusculis nervis secundariis utrinque paucis venis reticulatis, racemis terminalibus solitariis longis laxifloris, floribus nutantibus subsecundis coccineis, pedicellis pollicaribus basi bracteâ parvâ lineari stipatis, calycis laciniis linearibus integris stellatim patentibus tubum obconicum 2-3-plò superantibus, corollæ laciniis linearibus tubi angusti dimidium excedentibus, antheris glabris inferioribus 2 apice pilis liberis barbatis.—*Planchon.*

A decidedly twining habit, leaves with a twisted stalk and a long raceme of pendulous flowers, remarks M. Planchon, distinguish this from all the other Siphocampyls in cultivation. Mr. Linden, who introduced it to Europe, says that it was discovered by his collectors, Messrs. Funck and Schlim, several years ago, near Galipan, in the province of Caraccas, at the height of 5000 feet above the sea. Mr. Van Houtte adds that, like many other climbers, this will not flower till it has arrived at a considerable size : a fault, however, recompensed by the great quantity of flowers that follow the age of barrenness. It should be planted out in a warm conservatory, in a rich soil, and its stems trained to a trellis. It strikes from cuttings very unwillingly ; but it may be expected to be on sale by March, 1853.—*Flore des Serres.*

638. SENECIO CONCOLOR. *De Candolle.* A handsome greenhouse herbaceous plant. Native of the Cape of Good Hope. Flowers rich purple. Belongs to Composites. Introduced by Sir Charles Hulse, who received the seeds from Colonel G. Buller.

We have little doubt that this fine showy perennial is that which De Candolle meant by his *S. concolor*, from Tulbagh and the Kat River mountains, although in cultivation it scarcely produces any of the hairs to which his specific character points. It is evidently a near relation of the old *Senecio speciosus*. The root-leaves are spathulate, lanceolate, long-stalked, sinuated towards the base and toothed ; the leaves next above them are oblong and stalked ; the highest are sessile, and slightly stem-clasping and downy at the edges ; all are more or less incised. The stem grows about two feet high, and forms an open corymb scantily clothed with foliage. The flower-heads have a rich purple colour throughout, with a diameter of nearly two inches, most of which belongs to the rich purple ray. It requires to be treated like Cape Pelargoniums, grows freely in a mixture of loam, peat, and leaf mould, and is increased from seeds. It is a very handsome plant, in the way of a " Cineraria," and may prove useful for bedding out during the summer. Flowers in August and September.—*Journ. of Hort. Soc.*, vol. vii.

639. HOYA FRATERNA. *Blume.* A hothouse climbing plant from Java. Flowers buff-coloured. Belongs to Asclepiads. Introduced by Messrs. Veitch.

A very fine new and very distinct species of Hoya, first detected in Java by Blume, and since by Mr. Thomas Lobb, and sent by him to his employer Mr. Veitch, in whose stove at Exeter it has grown very vigorously, and yielded its very handsome flowers during a great part of the summer and autumn. Some of the leaves measure a foot in length : our coloured figure is taken from a portion of the plant yielding smaller foliage ; but these leaves are remarkable no less for their great size than they are for their firmness and thickness, and the very indistinct remote pinnated nerves, scarcely seen except when the leaf is held between the eye and the light, or when the leaves are dried for the herbarium ; then the shrinking of the parenchyma brings the veins more distinctly into view, and shows them to be pinnated, anastomosing, and slender. The petioles and costa beneath are peculiarly thick. The upper side of the corolla, disc excepted, is downy, or between silky and velvety, and of a pale yellowish buff-colour, but five stains or spots are seen radiating from the centre towards the sinuses, which are always wet and clammy, which clamminess appears to be due to a flow of honey from beneath each of the leaves of the crown or nectary, and give a rich brown tone of colour to the whole umbel of flowers. It was named *fraterna* by Blume, on account of its affinity to *H. coriacea*, from which it is however abundantly distinct. A climber, with terete stems and branches, rooting near the insertion of the petioles, bearing opposite leaves, on rather short but very thick petioles ; varying from six inches to a foot in length, singularly thick, and firmly fleshy, subcoriaceous, elliptical, very glabrous and even, the margins recurved, the apex rather acute, the base emarginate or subcordate, dark green and glossy above, pale and opaque beneath, where the midrib is very broad and prominent ; lateral veins scarcely at all visible except the leaf be held between the eye and the light, when they are seen to be pinnated, distant, slender, anastomosing towards the margin. Peduncle much shorter than the leaves, moderately stout, thickened at the base, bearing at the apex a dense umbel of rather large, brownish red flowers. Sepals five, oval, concave. Corolla rotate, pale buff, with five red brown blotches, five-lobed, the lobes triangular, silky, reflexed. Leaflets of the corona pale buff, rotundato-ovate, thick, fleshy, concave above, with a blood-red spot at the base, grooved beneath.—*Bot. Mag.*, t. 4684.

640. ALSTROMERIA PLANTAGINEA. *Martius.* A very fine herbaceous plant, with rich bell-shaped flowers of deep orange, lined with yellow, tipped with green, and spotted with dark brown bars. Native of Brazil. Belongs to Amaryllids. Introduced by M. de Jonghe. (Fig. 309.)

A. *plantaginea ;* herbacea 1—1½ pedalis flore excepto glaberrima, foliis ad apices ramorum sterilium confertis in ramis fertilibus plus minus inter se approximatis aversis lineari-lanceolatis (3—4 poll. longis) apice sphacelato acutiusculis margine integro pellucido lævibus 5—7-nerviis paginâ sursùm spectante lucidâ lætè viridi alterâ pallidiore supremis pseudo-verticillatis inæqualibus pedicellis multò brevioribus, umbellâ terminali 6—8-florâ, pedicellis (3-poll.

longis), sulcatis, floribus nutantibus, ovario subgloboso crassè sex-costato, perianthii parùm irregularis laciniis spathulatis, filamentis parum inæqualibus leviter incurvo-deflexis breviter exsertis, polline aureo, stylo staminibus subæquali glaberrimo trigono apice trifido divisuris linearibus stigmaticis.—*Planchon.*

This noble species was obtained some years since by M. de Jonghe of Brussells, through his collector Libon, who found it in the mountainous province of the Mines. It is a herbaceous plant, with simple erect stems, having no kind of tendency to twine, some terminated by a large umbel of flowers, such as are shewn in the accompanying cut, while others are merely stopped by a tuft of ribbed leaves. It requires exactly the same treatment as other Alstromerias.—*Planchon, in Flore des Serres.*

641. SOBRALIA CHLORANTHA. *Hooker.* A showy terrestrial Orchidaceous plant. Native of Brazil. Flowers yellow. Introduced by Lucombe, Pince, and Co.

Received in a flowering state from the stove of Messrs. Lucombe, Pince, and Co., in June, 1852. It was sent to them by Mr. Yates, from Para, in Brazil. The flowers are in general structure like those of Sobralia, but of a yellow colour, and with foliage more like that of some Cattleya, thick and leathery. Pœppig and Endlicher have a genus Cyathoglottis (*Nov. Gen. et Sp. Plant.*, etc., p. 55), which they distinguish from Sobralia by very slight characters, adding " *Sobraliæ* tamen proxime affine videtur," and which has yellow or white flowers : but the anther should be terminal, not, as here, attached to the middle lobe of a trifid apex to the column. In our plant, however, the lobes are shorter than in the red-flowered Sobralias, and the sepals as well as the petals are connivent and united for some length at the base. Whether the two genera be distinct or not, our species by no means accords either with *Cyathoglottis crocea* or *C. candida*, the only two described by Endlicher and Pœppig. With the root and base of the stem we are unacquainted. The portion sent to us is scarcely a span long including the leaves, and with no appearance of pseudobulb. The stem is about as thick as a goose-quill, nearly terete, covered for the most part with the long rather compressed sheathing bases of the leaves. Leaves two or three, very unequal in size ; the lowest of them half a foot long, the uppermost from one to two inches, resembling a bractea, all of them dark full green, oblong or elliptical-ovate, rather acute, subcoriaceous, fleshy, the margin a little recurved, the surface marked with a few, distant, parallel, longitudinal striæ. In a sterile plant sent us, the leaves are more nearly equal and more oblong. The flower is large, terminal, sessile, curved, of a uniform pale sulphur-coloured yellow. Ovary clavate, sessile, rising a little above the sheath of the upper or bracteal leaf. Sepals four inches long, erecto-connivent, acuminate, united for some little way above their base. Petals uniform with the sepals and of the same length, erecto-connivent. Lip erect, for the greater part of its length enclosed within the sepals and petals, large, longer than the perianth, broadly obovate, retuse, clawed at the base, the apex curved back and much waved : the disc faintly striated, with a slight elevation where the claw is set on, and below that two oblong, small, incurved scales or portions of the margin. Column clavate, curved, about two-thirds the length of the flower, yellow, deeper-coloured and plain in front ; the apex obscurely trifid, the lobes, especially the latter ones, short, obtuse ; the anther-case hemispherical, imbedded, as it were, within the lobes, and attached to the intermediate one.— *Bot. Mag.*, t. 4682.

642. MERIANIA KARSTENII. *Naudin.* (*aliàs* Meriania macrantha *Linden ;* aliàs Schwerinia superba *Karsten ;* alias Chastenæa longifolia *Naudin.*) A beautiful hothouse shrub, with rich crimson flowers. Native of the Caraccas. Belongs to Melastomads. Introduced by Mr. Linden.

The genus Meriania, which was dedicated by the Swedish Botanist Swartz to the memory of Sibylle de Merian, a Dutch lady who published a great work on the insects of Surinam, contains a small number of Melastomads inhabiting the West Indies and the intratropical Andes, all remarkable for the delicate venation of their leaves, and the brilliancy of their flowers. Karsten's genus Schwerinia certainly belongs to it, for the pretended distinction between the anthers of the two genera (two pores in one and one pore in the other) is too slight to possess real value. The species in question was found in the Caraccas by Mr. Linden in 1842, and forms No. 35 of his herbarium. It inhabits the middle mountain region among quantities of Thibaudias, Ternstrœmiads, Weinmannias, Myrtleblooms, Gesnerads and arborescent ferns, and like all such plants does best in a warm greenhouse.—*Planchon.* Mr. Van Houtte adds that it is a rival of *Pleroma elegans.* It is in fact a shrub with oval-lanceolate acuminate serrated 3-ribbed dark green leaves, and flowers as large as an apple-blossom, but with the peculiar colour of Lemonia.—*Flore des Serres.*

643. RHODODENDRON LOUIS PHILIPPE. A magnificent hardy hybrid, between *R. ponticum* and *R. arboreum*, with intensely crimson flowers.

This brilliant variety was obtained from seeds sown nearly ten years ago by M. Bertin, of Versailles, who also succeeded in raising at the same time two other very remarkable plants, viz., the Rhododendrons *Charles Truffaut* and *Madame Bertin.* It is said to be a variety of *Rhododendron arboreum,* and the brilliant colour of its flowers is in favour of this supposition ; in consequence, however, of the indiscriminate use of the word hybrid, which is often applied even by eminent persons to simple varieties produced by seed, we confess we have some hesitation in adopting the above opinion. The new Rhododendron is at all events hardy even at Paris ; it first flowered in 1846, but did not become generally known until last year, as M. Bertin was desirous of ascertaining that its characters were constant before it became an article of trade. The name by which the plant is designated commemorates at once benefits conferred and misfortunes suffered, and tends to excite feelings of gratitude and sympathy in the minds of those who were ever interested in the late king of the French. The following are the characters of this beautiful variety as given by M. Henzé :—A freely flowering shrub, having from its very base extremely ramified branches ; flowers developed early (April and May) ; leaves of an average size, oblong-lanceolate, smooth, mucronate, clear green above, paler below, petiole middling long, greenish. Flowers in an almost hemispherical and closely packed corymb. Flower-bud round, whole-coloured, with greenish scales ; peduncle moderately long, green in a house, and reddish in the open air. Calyx campanulate, moderately long, cup-shaped, with shallow divisions. Corolla moderately long, cup-shaped at the top with irregular very shallow rounded divisions scarcely undulating at the edges. General colour of a very brilliant lac red, set off by rich dark purple spots covering the upper divisions and a quarter of the lateral ones. Stamens with filaments reddish at the base and scarcely projecting beyond the corolla ; anthers darker in colour, spotted, yellow ; style reddish, longer than the stamens ; stigma brown.—*Planchon, in Revue Horticole,* 1852, p. 361, fig. 19. Certainly, according to the figure, a very fine variety, with all the brilliancy of the best states of *R. arboreum.*

644. PHALÆNOPSIS INTERMEDIA. A very fine stove epiphyte. Flowers white and deep rose. Introduced by Messrs. Veitch & Co. (Fig. 310.)

P. intermedia; petalis latè rhombeis acutis, labelli lobis lateralibus cuneatis obtusangulis intermedio ovato apice bicirrhoso.

It is not improbable that this beautiful plant is a natural mule between *P. amabilis* and *rosea*. It agrees with the former in foliage and in the tendrils of the lip ; with the latter in colour, in the acuteness of its petals, and in the peculiar form of the middle lobe of the lip. The short description will enable it to be easily recognised. Axis of inflorescence deep brownish purple, bearing flowers half way in size between *P. amabilis* and *rosea*. Sepals pure white, concave, oblong, acute. Petals much larger, lozenge-shaped, acute, pure white with a few minute speckles at the base. Lip three-lobed ; the lateral divisions erect, wedge-shaped, with rounded angles, violet with a few crimson spots and dots ; the middle division ovate, deep crimson, with the point separated into two short tendrils. Crest at the junction of the lobes of the lip nearly square, depressed in the middle, deep yellow with crimson dots.

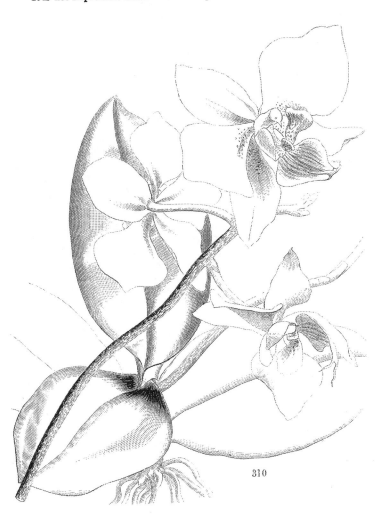

310

645. ROGIERA CORDATA. *Planchon.* (*aliàs* Rondeletia cordata *Bentham*.) A beautiful stove shrub, with rich pale rose-coloured cymes of flowers. Native of Guatemala. Belongs to Cinchonads. Introduced by Mr. Van Houtte.

A beautiful shrub in the way of *Rogiera amœna*, figured in our first volume among the Gleanings, no. 194, fig. 95. Its leaves are distinctly heart-shaped, and thus are readily known from the others. The colour is said to be brighter, and the eye of a more clear yellow.—*Planchon, in Flore des Serres.*

646. RUBUS JAPONICUS. *Veitch.* A hardy shrub, with broad bright green leaves, white flowers, and yellow fruit. Native of Japan. Belongs to Roseworts. Introduced by Messrs. Veitch and Co.

R. japonicus; erectus, inermis, glaberrimus, foliis simplicibus altè cordatis palmatis lobis duplicato-serratis, stipulis integris lineari-oblongis acutis, floribus 2—3 terminalibus pedunculis calycibusque glanduloso-tomentosis.

Messrs. Veitch and Co. received this from Mr. T. Lobb, who found it growing in the Botanic Garden at Buitenzorg, where it had been introduced from Japan. We do not find it described. It forms a bush, with the habit of *R. nutkanus*, erect, with no prickles whatever on the stem. The leaves are as large as those of a Sycamore, thin, dark bright green, shining and handsome ; the uppermost are smaller, and gradually become three-lobed. The flowers appear two or three together at the end of the branches, are white and inconspicuous, but they are succeeded by beautiful yellow raspberries, rather pleasant to the taste, and as large as those of the " Yellow Antwerp." The shrub seems to be hardy ; in Messrs. Veitch's nursery it was planted in front of a low wall, where it proves to be almost evergreen.

647. ECHINOPSIS CRISTATA. *Salm-Dyck.* (*aliàs* Echinocactus obrepandus *Salm-Dyck.*)   A very fine succulent plant, with large straw-coloured flowers.   Native of Bolivia.

This, as well as the purple-flowered variety of it, were imported by Mr. Bridges from Bolivia (not Chili, as stated by Mr. Smith in *Bot. Mag.*, under t. 4521).   The latter is already figured in the plate just cited, and we scarcely know which is the more striking of the two.   The purple-flowered variety has the advantage in the colour of the flower, but the present kind produces the largest blossoms ; the petals are broader in proportion to their length, a cream-white gradually passing into the greenish purple of the outer sepals.   The spines in the present variety are more slender, less curved, of a paler colour, but tipped with a darker brown.   In other respects the two plants correspond.—*Bot. Mag.*, t. 4687.

648. HEDYCHIUM FLAVESCENS. *Loddiges.* (*aliàs* H. Roxburghii *Siebold.*)   A handsome and fragrant stove plant, native of India.   Flowers pale yellow, in August.   Belongs to Gingerworts. (Fig. 311.)

A stout plant, about four feet high, with a great fleshy rhizome.   Leaves about fifteen inches long by five inches broad, covered on the under side with long silky hairs.   Flower-spike erect, a foot long, covered with brown hairs. Outer bracts rather distant, two inches long, with a short leafy revolute point, and closely covered with rusty hairs at the edges ; rolled round a very short spike of five flowers, surrounded by membranous, nearly-smooth bractlets.   Ovary and long tubular calyx shaggy with brown hairs.   Tube of the corolla smooth, slender, four inches long ; its three outer petals linear and revolute ; of the pale yellow three inner, the lateral are unguiculate, spathulate-lanceolate, acuminate, slightly toothed ; the lip is unguiculate, deeply two-parted with half-oval divisions, about half the length of the bright orange-coloured filament.   The flowers emit a very agreeable spicy fragrance.   It requires to be potted in a rich loamy soil and to be placed in the dampest part of the stove while in a growing state ; afterwards it should be removed to a cooler and drier place to bloom ; after flowering it should be dried gradually, and rested for about a month.   It blossoms in September.   It is a very fragrant plant, and rather showy, but remains only a short time in bloom.   There is no doubt about its being the *H. flavescens* of the Botanical Cabinet ; but I should have thought it to be also *H. villosum* of Dr. Wallich, if that plant had not been described as having five linear petals, whereas here three only are linear and two broad spathulate-lanceolate.—*Journ. of Hort. Soc.*, vol. vii.

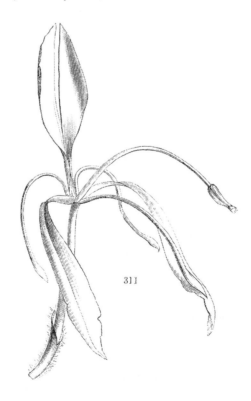

311

PLATE 106.

L.Constans del.& zinc.

Printed by C.F.Cheffins, London.

[Plate 106.]

THE CHINESE ALTHÆA FRUTEX.

(HIBISCUS SYRIACUS; VAR. CHINENSIS.)

———◆———

A beautiful Stove Shrub, native of China, *belonging to* Mallowworts.

———————

Hibiscus syriacus : *Linnæus.*

═══════════

THE common *Althæa frutex* is said upon no very good authority to be a native of Palæstine, and even of Carniolia ; but it does not appear to have been known to the Greeks, and Forskähl expressly states that it is a garden plant in Egypt. *Colitur in hortis Ægypti ; floribus splendidis ; aut totis violaceis, vel albis, basi rubris.* (Fl. ægypt. arab., p. 125.) Its real country must in truth be regarded as unknown ; it however appears to be very common in the East of Asia, but always cultivated. Thunberg tells us that it is grown every where in Japan for live fences, and that it is the *Kin* of Kæmpfer. Of this *Kin* the latter author tells us that it is also called *Mu Kunge,* that it is cultivated, and has in one state single flowers, blue shading into purple, *flore in purpureum cæruleo,* in another state double tinged with blue, *cæruleato,* with dense crisp petals, but neither style nor stamens. (Amœn. exot. 858.)

One of these forms is now before the reader in the accompanying plate, drawn in the garden of the Horticultural Society, where it had been raised from seeds, presented to the Society by John Reeves, Esq., in June, 1844, under the name of Koorkun Vellory.

The Editor of the Society's Journal speaks thus of the plant itself :—

" I think there can be no doubt that this, although certainly Chinese, is a mere variety, and not a well marked one, of *Hibiscus syriacus.* It has large violet flowers, with a crimson eye, and its leaves are larger, thinner, and more smooth than in the shrub out of doors, owing, perhaps, to having been grown in a stove. But the last circumstance is evidently unimportant, for in Mr. Fortune's wild specimens now before us, the leaf-stalks are perfectly shaggy. This traveller found it forming a shrub eight to twelve feet high, with light 'blue' flowers, in the hedges and on hill-sides on Poo-too-san, and other islands.

"When growing in a stove, with the same kind of treatment as is required by the well-known *Hibiscus rosa-sinensis*—that is to say, if grown in a mixture of sandy loam, peat, and leaf-mould, it forms a very handsome shrub, flowering in July and August."

PLATE IC

L.Constans del.& zinc.

Printed by C.F. Cheffins, Lond

[PLATE 107.]

THE CALISAYA BARK-PLANT.

(CINCHONA CALISAYA.)

◆

A fragrant Hothouse Shrub, native of BOLIVIA, *belonging to the Natural Order of* CINCHONADS.

═══════════════════════

𝕾𝖕𝖊𝖈𝖎𝖋𝖎𝖈 𝕮𝖍𝖆𝖗𝖆𝖈𝖙𝖊𝖗.

THE CALISAYA BARK-PLANT. Leaves oblong or lanceolate-obovate, obtuse, narrower at the base, seldom sharp at both ends, smooth and shining or downy on the underside, with pits in the axils of the veins. Filaments not half so long as the anthers. Capsule ovate, scarcely so long as the flowers. Seeds finely and closely fringed with teeth at the edge.

CINCHONA *CALISAYA ;* foliis oblongis v. lanceolato-obovatis obtusis basi attenuatis rariùs utrinque acutis glabratis nitidis v. subtus pubescentibus in axillis venarum scrobiculatis, filamentis quam dimidia antherâ plerumque brevioribus, capsulâ ovatâ flores longitudine vix æquante, seminibus margine crebrè fimbriato denticulatis. — *Weddell.*

Cinchona Calisaya : *Weddell, Hist. Nat. des Quinquinas,* p. 30, tt. 3 and 4 ; *Journal of Hort. Soc.,* vol. vi. p. 272.

═══════════════════════

W͟E owe our knowledge of this important plant to one of the boldest and best of the naturalists employed by the French Government. Dr. Weddell, an English Botanist, attached to the mission of M. de Castelnau, succeeded, among innumerable difficulties, in reaching the country where the Calisaya, the most precious of the kinds of Cinchona, or Peruvian Barks, is produced. He brought seeds to Europe; and from some of them, obtained from the Jardin des Plantes of Paris through the friendly assistance of J. B. Pentland, Esq., the Horticultural Society raised the plant whose flowers are now represented. From the very full account of it in the Society's Journal we make as many extracts as our space will permit.

"The leaves are oblong, obtuse, pale dull green, tapering gradually into the petiole, which is red, as well as the midrib itself; at the back of the leaf, in the axil of each principal vein, is a small excavation closed up by hairs. The stipules, which fall off very early, are a pair of oblong, erect, blunt, smooth plates. The flowers appear in panicles at the ends of the lateral shoots, are of a pale pink colour before expansion, almost white when fully open, and emit a most agreeable weak balsamic fragrance. The calyx is a small superior five-toothed cup, covered with fine close down like the branches of the panicle. The corolla has a cylindrical tube about half an inch long, and a reflexed five-lobed limb, copiously fringed with long transparent club-shaped hairs. The stamens are five, and can just be seen when looking down into the tube of the corolla.

Dr. Weddell, in his Natural History of the Quinquinas, states that :—

" From this species is obtained the most precious of the Jesuit's barks used in medicine, employed from time immemorial in trade under the name of Calisaya bark, but whose origin was wholly unknown till now.

" I have already observed that this tree has hitherto been only found in Peru, in the southern part of the province of Carabaya. The results at which I have arrived in endeavouring to determine exactly the limits of the region it occupies seem curious enough to be noted in this place. Thus, after having studied the plant in all the ancient province of Yungas in La Paz, to the north of 17° S. lat., I followed it into that of Larecaja or Sorata, thence into Caupolican or Apolobamba, the place of its first discovery ; and all my care has failed in enabling me to find it north of those points. An imaginary barrier exists then beyond which the plant will not go, notwithstanding that the neighbouring valleys appeared to be of the very same nature ; a fact that can scarcely be explained, unless upon the supposition that peculiarities do exist in the most southern valleys of Carabaya which are wanting in the north ; and this may possibly be owing to the manner in which the rivers are distributed. I believe, in fact, that I am justified in referring those of the district in question to a particular system, possibly dependent upon the Bolivian system, and that those in the other parts of the province lose themselves on the contrary by the N. of Peru, in the Upper Amazon. This unexplained attachment which certain plants manifest for natural regions, and especially for valleys, is by no means without example ; and now that Geographical Botany is obtaining serious attention, science will be enriched more and more with analogous facts.

" The great reputation of the Quinquina Calisaya has caused such a demand for it, that it will certainly some day disappear completely from commerce, and we shall be obliged to be content with other sorts now despised. It has already disappeared around inhabited places, except in the form of a bush ; and if by mere chance a small tree has remained unobserved in the midst of a forest, its head no sooner becomes visible than the hatchet brings it down. For my own part, when I have wished to see the species in all its vigour, it has been necessary to pass long days on foot in the forests, to penetrate them by paths which were scarcely passable, and to undergo some of the fatigues which are the ordinary lot of the poor Cascarilleros."

Its native station was found by this enterprising traveller to be on the slopes and precipices of mountains as high as 4500 or 5400 feet in the hottest valleys of Bolivia and Southern Peru, in forests between 13° and 16° 30′ S. lat., and 68°—72° W. long., in the Bolivian provinces of Enquisivi, Yungas, Larecaja, and Caupolican, and in Carabaya in Peru.

This plant has been found to require very peculiar management. Mr. George Gordon, under whose care it flowered in the Society's Garden, explains at length in the " Journal " in what way the specimen was treated which bloomed so abundantly in the Society's stove, and the reader is referred thither for information.

PLATE 108

L.Constans del.& zinc.

Printed by C.F.Cheffins, London.

[PLATE 108.]

THE SPLENDID ÆSCHYNANTH.

(ÆSCHYNANTHUS SPLENDIDUS.)

———◆———

A magnificent Stove Plant, of GARDEN ORIGIN, *belonging to the Natural Order of* GESNERADS.

OF this most beautiful thing we have the following account from Messrs. Lucombe, Pince, and Co., of Exeter, who raised it.

"We have very great pleasure in sending you a cut specimen of our new *Æschynanthus splendidus,* which we think you will admire. It is a hybrid produced from *Æ. speciosus* impregnated with *Æ. grandiflorus,* and possesses the brilliancy of colour and hardy constitution of the male, whilst it also fully partakes of the many good qualities of the other parent.

"It is easily cultivated, producing a long succession of large umbels of brilliant coloured flowers, and requires much less heat than many other Æschynanths, a circumstance easily accounted for by the fact that *Æ. grandiflorus* has been frequently wintered by us in a cold pit, into which frost has sometimes penetrated. A figure of *Æ. splendidus* has been published in a contemporary in December last, but it did not by any means do justice to the subject, and the specimen I now send is better even than that from which the drawing was made. In no respect has this fine hybrid had that publicity given to it which such a plant merits."—*Exeter, Sept.* 7, 1852.

At a later period many small plants were exhibited to the Horticultural Society, for the purpose of showing how abundantly they blossom even in the youngest stage. They formed a brilliant circle, of which it is no exaggeration to say, that all other colours became pale when contrasted with theirs.

GLEANINGS AND ORIGINAL MEMORANDA.

649. CŒLOGYNE CRISTATA. *Lindley.* A beautiful Orchidaceous epiphyte, from the North of India. Flowers large, pure white, with numerous yellow fringes on the lip. (Fig. 312.)

One of the most striking of the white-flowered Indian Orchids. It forms oblong or ovate two-leaved pseudobulbs upon a hard scaly rhizome. The leaves are lanceolate, tough and flaccid, with some waviness at the edge. The flowers appear in long drooping imbricated spikes ; in the beginning they are concealed by brown dry spathes, which afterwards sheathe the ovary and its stalk. When expanded they form a pendent raceme, consisting of from four to six, each fully four inches in diameter when fully expanded. The sepals and petals are pure white, lanceolate, wavy, and acuminate. The lip, which is also white, is concave, and three-lobed ; the lateral lobes half-oblong, truncate at the upper end, and somewhat wider than the transverse roundish three-toothed middle lobe. Along the middle run five parallel veins covered by delicate yellow glandular fringes ; at the base of each of the three central is a wavy plate, and at the upper end of the two which stand on each side the middle vein is another solid plate terminating abruptly in front and more or less toothed. The very fine specimen, of which our cut represents a portion, flowered at Chatsworth in March, 1850. In that specimen the pseudobulbs were fully three inches long ; but they are usually much smaller.

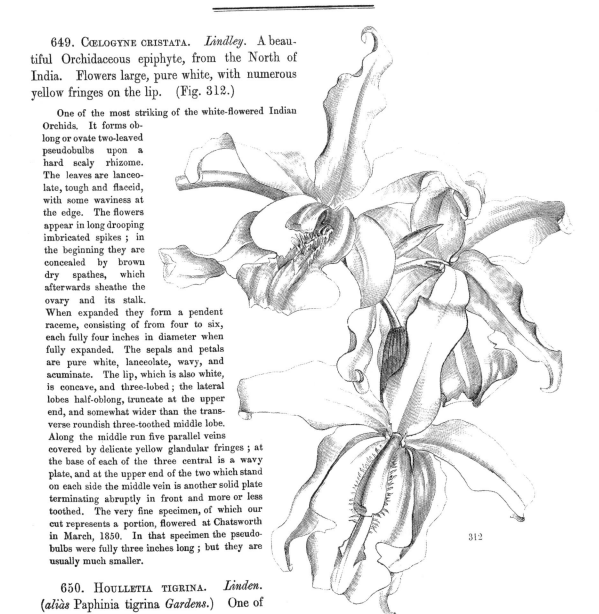

312

650. HOULLETIA TIGRINA. *Linden.* (*aliàs* Paphinia tigrina *Gardens.*) One of

the finest of all epiphytes. Flowers rich yellow daggled all over with crimson. Native of New Grenada. Belongs to the Order of Orchids.

H. *tigrina*; scapo decurvo, sepalis altè connatis, petalis acutissimè trilobis, labelli epichilio sessili ovato obtuso apiculato hastato versus basin verrucoso angulis posticis acuminatis, mesochilio apice carnoso in tuberculum infundibulare elevato cirrhis ascendentibus falcatis columnâ brevioribus, hypochilio carnosissimo basi excavato semibiloculari.

Wild on the ground in New Grenada, in the province of Ocaña, growing in forests of Weinmannia, where it was found by Mr. Schlim, one of Mr. Linden's collectors, in June and July, 1851, at an elevation of 4800 feet. It has been our good fortune to see many beautiful Orchids, and after becoming acquainted with *Phalænopsis amabilis*, *Vanda cærulea*, and a few others of that class, we had supposed that nothing finer remained to reward the traveller. But here we have a novelty which, to say the least of it, must be ranked fully on a level with those glorious species. It has just flowered with Mr. Linden, in his establishment at Luxembourg, and we have now before us some recent flowers and a magnificent coloured figure by De Tollenaere, prepared for publication in the present work ; but as this is the last number of our " Flower Garden " which will appear, a brief description is all we can give. The leaves are broad, plaited, erect, almost two feet long, and look like those of a vigorous Stanhopea. The stem appears to be about as thick as a swan's quill, greenish red, slightly dotted with brown. The flowers are four inches in diameter, and of the firm texture of Stanhopeas. The sepals are oblong, concave, straw-colour, very richly mottled and variegated with deep rose. The petals are one inch and three-quarters long, very acute, with a strong sharp-pointed lobe on each side, brilliant yellow variegated with rich crimson in the same way as the sepals. The lip consists of a broad fleshy oblong stalk and a flat spade-shaped blade ; the latter is yellowish at the point, otherwise whitish speckled all over with crimson ; the stalk is richly marked with cross bands of blood-red, and has on either side a process shaped like a scythe-blade which rises up in the direction of the anther. The column is dull yellow speckled with purple. After such an account it is only necessary to say that Mr. Linden has this noble plant on sale, and that the character of Houlletia as a genus distinct from Stanhopea is placed in jeopardy.

We avail ourselves of the present opportunity of mentioning that Mr. Linden also possesses another Houlletia, having much the habit of *H. Brockelhurstii*, and like it remarkable for its fragrance. It also grows in New Grenada, in the province of Ocaña, where it was found by Mr. Schlim in May, 1851, on the borders of rivulets. Its extremely aromatic odour discovered its presence at a considerable distance, on which account Mr. Linden calls it *H. odoratissima*. The pseudobulbs are described as resembling those of the Brazilian Houlletia, but being more blunt. The flowers are brick-red, with the lip and column white. The following technical definition will explain to the botanist how much it differs from the last :—

651. H. *odoratissima* Linden ; scapo stricto, sepalis liberis, petalis sepalis conformibus indivisis, labelli epichilio unguicu-lato ovato obtuso subsagittato undique intra marginem verrucoso angulis posticis obtusis, mesochilio dente longo linguiformi apice acuto cirrhis ascendentibus falcatis columnâ brevioribus, hypochilio appendice pedicellatâ cyathiformi aucto.

652. BEGONIA XANTHINA. *Hooker.* A noble hothouse species of Begoniad, native of Bootan. Flowers deep yellow.

A very beautiful new Begonia cultivated by Mr. Nuttall in his stove at Rainhill, near Preston, Lincolnshire, where it was raised from roots sent in 1850 from Bootan, by his nephew, Mr. Booth. It is remarkable for the large, full, almost golden-yellow flowers, tinged with red at the back, which contrast well with the ample foliage of a deep glossy green above, and with the fine red of the petioles, peduncles (shaggy, with scale-like hairs), and underside of the leaf. It flowered in July, 1852. Root a short, thick, horizontal, fleshy rhizoma, shaggy with scaly hairs at the setting on of the petioles, and bearing fibrous radicles below. Stem none. Leaves ample, six inches to a span or more long, obliquely (inequilaterally) cordato-ovate, shortly acuminated, more or less sinuated, the margin denticulated, subciliate, penninerved and reticulately veined, of a deep full glossy green and glabrous above, beneath red, with the nerves prominent, the chief ones and costa hispid. Petioles thick, fleshy, terete, bright red, about a span long, clustered from the apex of the rhizoma, and there having large, ovate, submembranaceous, coloured stipules ; their peduncles are crinite, with shaggy patent hairs, almost scaly and reflexed below. Peduncles twice as long as the petiole, and resembling it, but glabrous above, bearing a many-flowered corymb at the extremity. Flowers deep full yellow, drooping, often springing three from one point, in which case two are male flowers, and one is female. Male flower much the largest, of four spreading sepals, of which three are oblong-obovate, and the fourth rotundate, larger and more concave, tinged with red at the back. Stamens very numerous, forming a compact, globose, yellow head. Female flower small, of six nearly orbicular, concave, erect petals, tinged with red at the back. Fruit greenish, tinged with red, three-winged, two of the wings short and equal, the third is remarkably elongated horizontally, into a sort of broad blunt beak, and is striated.—*Bot. Mag.*, t. 4683.

653. SPHÆRALCEA NUTANS. *Scheidweiler.* A coarse purple-flowered greenhouse shrub. Native of Guatemala. Belongs to Mallowworts. Introduced by Mr. Van Houtte. (Fig. 313.)

This forms a branching shrub, with the habit of an Abutilon or Hibiscus. The leaves are palmate, long-stalked, and divided into five very acute serrated lobes. A thick felt of stellate hairs, which cover almost every part and especially the leaves, gives them a grey unpleasant appearance. But, on the other hand, the great carmine-rose flowers drooping gracefully from the end of a long peduncle produce a sufficiently brilliant effect. It is not certain that the

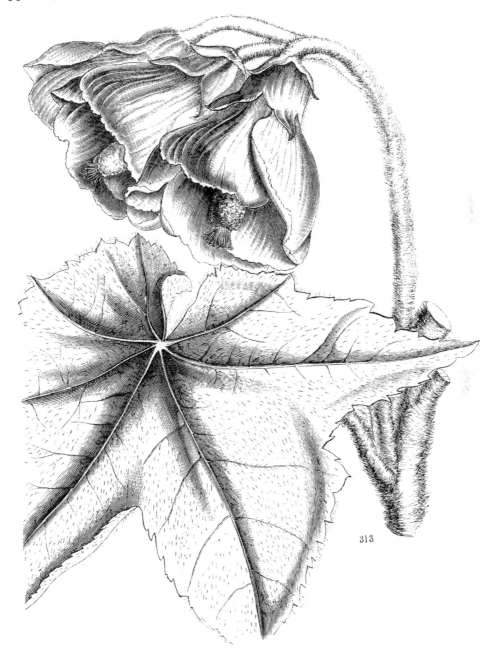

313

plant comes from Guatemala, but it is supposed to do so by Mr. Van Houtte. The genus *Sphæralcea* differs from Malva in its dehiscent carpels containing three ovules instead of one ; that such is the structure of the plant now described rests upon the authority of Prof. Scheidweiler M. Planchon not having had an opportunity of verifying the fact.— *Flore des Serres*, 726.

654. Odontoglossum Pescatorei. *Linden.*

It now appears that this beautiful plant, figured at Plate 90, had been previously described by Mr. H. G. Reichenbach under the name of *Odontoglossum nobile*, as has been suggested in the *Folia Orchidacea*. It was however impossible to recognise the description, in consequence of the misprints which it contained ; and we question whether, under such circumstances, Mr. Linden's name can be disturbed.

655. Mormodes speciosum. *Linden.* A beautiful stove epiphyte, from Ocaña. Flowers deep yellow richly speckled with cinnamon. Introduced by Mr. Linden.

M. *speciosum ;* sepalis petalisque lanceolatis, labelli tripartiti glabri laciniis lateralibus ovatis obtusis intermediâ acuminatâ multò brevioribus.

A very fine species found by Mr. Schlim in New Grenada in the province of Ocaña, at the elevation of 4800 feet, in August, 1852. The appearance is that of the genus generally ; the flowers are three inches in diameter, of a deep golden-yellow, speckled all over even to the tip of the column with the rich cinnamon-red. The points of the lip are deep purple. This has just flowered at Luxembourg, with Mr. Linden.

656. Sophronites.

Mr. H. G. Reichenbach has pointed out to us that the false name of *Sophronitis nutans* ascribed to him by an accidental transposition of type, at No. 472, is really chargeable upon Hoffmannsegg ; and that the name of *S. Hoffmannseggii*, another false name, should be placed to his father's account.

657. Leptosiphon luteum. *Bentham.* (*aliàs* Gilia lutea *Steudel.*) A Californian hardy annual, with gay yellow flowers. Belongs to Polemoniads. Introduced by Messrs. Veitch & Co. (Fig. 314.)

This brilliant little plant was long since discovered by Douglas, but has only recently been introduced by Messrs. Veitch. With the habit of the other species now familiar in gardens it joins very bright yellow flowers, which in one variety are as pale as a lemon, in another as dark as an orange. It is perfectly hardy, and demands the same treatment as *L. androsaceus.*

314

INDEX OF VOLUME III.

[*Plate* signifies the coloured representations ; *No.* the number of the Gleanings and Memoranda ; *fig.* the woodcuts.]

THE END.

BRADBURY AND EVANS, PRINTERS, WHITEFRIARS.